Global Warming

& THE Greenhouse Effect

Grades 7–8

(Can be extended to Grades 9 and 10)

Skills

Observing, Measuring, Recording Data, Interpreting Graphs, Experimenting, Drawing Conclusions, Synthesizing Information, Role-Playing, Using Simulation Games, Problem Solving, Brainstorming Solutions, Critical Thinking

Concepts

Atmosphere, Visible and Infrared Photons, Greenhouse Effect, Sources of Carbon Dioxide, Climate and Weather, Effects of Climate Change, Molecular Model of Heat, Interaction of Energy and Matter

Themes

Systems and Interactions, Models and Simulations, Patterns of Change, Stability, Matter, Energy, Evolution, Structure, Scale, Diversity and Unity

Mathematics Strands

Number, Measurement, Logic and Language, Statistics and Probability, Function, Algebra

Nature of Science and Mathematics

Scientific Community, Science and Technology, Changing Nature of Facts and Theories, Creativity and Constraints, Theory-Based and Testable, Cooperative Efforts, Objectivity and Ethics, Real-Life Applications, Interdisciplinary

by

Colin Hocking
Cary Sneider
John Erickson
Richard Golden

Great Explorations in Math and Science (GEMS)
Lawrence Hall of Science
University of California at Berkeley

Lawrence Hall of Science
Chairman: Glenn T. Seaborg
Director: Ian Carmichael

Initial support for the origination and publication of the GEMS series was provided by the A.W. Mellon Foundation and the Carnegie Corporation of New York. GEMS has received support from the McDonnell-Douglas Foundation and the McDonnell-Douglas Employees Community Fund, the Hewlett Packard Company Foundation, and the people at Chevron USA. GEMS gratefully acknowledges the contribution of word processing equipment from Apple Computer, Inc. This support does not imply responsibility for statements or views expressed in publications of the GEMS program.

Under a grant from the National Science Foundation, GEMS Leader's Workshops have been held across the country. For further information on GEMS leadership opportunities, or to receive a publication brochure and the *GEMS Network News*, please contact GEMS at the address and phone number provided.

COMMENTS WELCOME

Great Explorations in Math and Science (GEMS) is an ongoing curriculum development project. GEMS guides are revised periodically, to incorporate teacher comments and new approaches. We welcome your criticisms, suggestions, helpful hints, and any anecdotes about your experience presenting GEMS activities. Your suggestions will be reviewed each time a GEMS guide is revised. Please send your comments to:

University of California, Berkeley
GEMS Revisions
Lawrence Hall of Science # 5200
Berkeley, CA 94720-5200

Our phone number is (510) 642-7771.
Our fax number is (510) 643-0309.

Visit the GEMS Web site at
www.lhsgems.org or e-mail us at
gems@uclink4.berkeley.edu

GEMS STAFF

Principal Investigator
Glenn T. Seaborg
Director
Jacqueline Barber
Associate Director
Lincoln Bergman
Associate Director/Principal Editor
Lincoln Bergman
Science Curriculum Specialist
Cary Sneider
Mathematics Curriculum Specialist
Jaine Kopp
GEMS Sites and Centers Coordinator
Carolyn Willard
GEMS Workshop Coordinator
Laura Tucker
GEMS Workshop Administrator
Terry Cort
Staff Development Specialists
Lynn Barakos, Katharine Barrett, Kevin Beals, Ellen Blinderman, Beatrice Boffen, Gigi Dornfest, John Erickson, Stan Fukunaga, Philip Gonsalves, Cathy Larripa, Laura Lowell, Linda Lipner, Debra Sutter
Distribution Coordinator
Karen Milligan
Distribution Representative
Felicia Roston
Shipping Assistants
Ben Arreguy, George Kasarjian
Senior Editor
Carl Babcock
Editor
Florence Stone
Public Information Representative
Gerri Ginsburg
Principal Publications Coordinator
Kay Fairwell
Art Director
Lisa Haderlie Baker
Designers
Carol Bevilacqua, Rose Craig, Lisa Klofkorn
Staff Assistants
Kasia Bukowinski, Larry Gates, Nick Huynh, Steve Lim, Nancy Lin, Michelle Mahogany, Karla Penuelas, Alisa Sramala

Printed in the United States of America. International Standard Book Number (ISBN):0-924886-65-X.

Great Explorations in Math and Science (GEMS) Program

The Lawrence Hall of Science (LHS) is a public science center of the University of California at Berkeley. LHS offers a full program of activities for the public, including workshops and classes, exhibits, films, lectures, and special events. LHS is also a center for teacher education and curriculum research and development.

Over the years, LHS staff developed a multitude of activities, assembly programs, classes, and interactive exhibits. These programs have proven to be successful at LHS and should be useful to schools, other science centers, museums, and community groups.

A number of these guided-discovery activities are published under the Great Explorations in Math and Science (GEMS) title, after an extensive refinement process that includes classroom testing, ensuring the use of easy-to-obtain materials, with carefully written step-by-step instructions and background information to allow presentation by teachers without special background in mathematics or science.

Contributing Authors

Jacqueline Barber
Katharine Barrett
Kevin Beals
Lincoln Bergman
Beverly Braxton
Kevin Cuff
Linda De Lucchi
Gigi Dornfest
Jean Echols
John Erickson
Philip Gonsalves
Jan M. Goodman
Alan Gould
Catherine Halversen
Kimi Hosoume
Sue Jagoda
Jaine Kopp
Linda Lipner
Larry Malone
Cary I. Sneider
Craig Strang
Debra Sutter
Carolyn Willard

Global Warming & the Greenhouse Effect

Illustrations
Rose Craig
Lisa Klofkorn
Carol Bevilacqua

Photos
Richard Hoyt
Colin Hocking
Lincoln Bergman

Acknowledgments

The first person to take the initiative to develop this series of educational activities about global warming and the greenhouse effect was Richard Golden. As Science Coordinator in New Rochelle, New York, Richard had coordinated field test trials for a great many of the teacher's guides in the GEMS series. In 1987, Richard moved to California where he began to work more closely with the GEMS staff in developing activities related to global warming.

Some of the ideas that led to this guide were first discussed at a conference of scientists, curriculum developers, school administrators, science teachers, and social studies teachers held at the Science Education Center, Lawrence Livermore National Laboratory. The conference was held on March 1–3, 1988, with cooperation from the Lawrence Hall of Science, the Oak Ridge National Laboratory, Oak Ridge Associated Universities, the Lawrence Berkeley Laboratory, with support from the U.S. Department of Energy.

Some of the key activities took shape in fall 1988, when Richard Golden and Cary Sneider undertook to develop a high school unit on global warming as part of a teaching methods course. The course was developed by the Education Department of the Lawrence Hall of Science in cooperation with Mills College of Oakland, California, under a grant from the National Science Foundation. Students in the course were men and women who were changing careers from scientific and technical fields into teaching. These participants, and the other faculty members from Mills College and the Lawrence Hall of Science, provided valuable criticism throughout the development process.

Early in 1990, our team expanded to include two experienced curriculum developers: Colin Hocking, from Australia, and John Erickson, from LHS. Colin worked on this unit while on leave from the Centre for Communications and Applied Science, Holmesglen College of Technical and Further Education, Melbourne, Australia. Additional grants from the Carnegie Corporation of New York, and the U.S. Department of Energy's Center for Science and Engineering Education at the Lawrence Berkeley Laboratory, allowed us to involve Colin and John in developing a new version of the unit for middle school students, and to rigorously test these instructional materials at several sites in the San Francisco Bay Area. We are indebted to the many teachers, listed on page vi, who assisted with these classroom trials, and to Gabriel Grinder and Chelsea Eis, the first two children to trial test the use of balloons for collecting and testing gas samples for the presence of carbon dioxide.

The "Effects Wheel" activity in Session 7 was based on the work of Sue Lewis and the McClintock Collective, a group of teachers working on gender and science in Melbourne, Australia. Noua's story was inspired by the work on guided fantasies by Serena Everill and the Social Biology

Resource Centre in Melbourne, Australia. Ideas for the "World Conference on Global Warming" were drawn from independent discussions with members of the San Francisco Bay Area Chapter of the Sierra Club, the United Nations Association of the U.S.A., and the Climate Protection Institute of Oakland,

California. The "The Great Global Warming Limerick Debate" poem is by GEMS Principal Editor Lincoln Bergman.

We are indebted to Dr. Stephen H. Schneider, professor in the Department of Biological Science and Senior Fellow at the Institute for International Studies at Stanford University, and Dr. Michael McCracken, director of the Office of the United States Global Change Research Program, for reviewing the manuscript of this teacher's guide for scientific accuracy. Their expertise in this complex and controversial field is very much appreciated. Of course, final responsibility for accuracy and factual interpretation rests with the authors.

Edna DeVore, John Erickson, and Chris Harper of the LHS Physics and Astronomy Department made numerous excellent suggestions for revisions of this guide.

Reviewers

We want to thank the following educators who reviewed, tested, or coordinated the reviewing of this GEMS teacher's guide in manuscript form. Their critical comments and recommendations contributed significantly to this publication. Their participation does not necessarily imply endorsement of the GEMS program.

CALIFORNIA

Albany Middle School
Albany
Susan Butsch
Robin Davis
Jack McFarland
Janet Obata Teel

Bancroft Junior High School
San Leandro
David Collins
Catherine Heck
Paul Mason Hynds
Ann Mosle

Bancroft Middle School
San Leandro
Stephen Rutherford (Trial Test Coordinator)

Columbus Intermediate School
Berkeley
Joy Osborn
Phoebe A. Tanner

Martin Luther King Junior High School
Berkeley
Jaine Gilbert

Piedmont Middle School
Piedmont
Kerri Lubin (Trial Test Coordinator)
Lynn Allen
Marianne Gielow
Camilla Thayer
Larry Zedaker

St. Paul's Episcopal School
Oakland
Karen Ginsberg Beroldo
Susan B. Porter

WASHINGTON

Whatcom Middle School
Bellingham
Ann L. Babcock

The following individuals also made important contributions to the development of these units, as part of a mid-career teaching credential program at Mills College in Oakland, California—see Acknowledgments on page iv for further details.

MILLS COLLEGE
DEPARTMENT OF EDUCATION
Jane Bowyer
Betty Karplus
Claire Smith

1988-89 Class Members
John Cypher
Marianne Gielow
Margaret Hellweg
Dick Holmquist
Yvette McCullough
Annie Peterson
Simone Saunders
Alfreda Stephens-Surge
Robert Stewart
Jennifer Wilson

1989-90 Class Members
Lynn Allan
Lisa Aronow
Richard Bell
Bob Brewer
Lejla Cyr
Steven Eiger
Vincent Haskell
Peter Hollingsworth
Jefferey Knoth
Dennis Kohimann
Karen Kyker
Sandi Metaxas
Eleanor Rasmussen
Wendy Struhl
Timothy Tisher
Jeff Winemiller

Contents

Why Should My Students Study Global Warming?

The topic of global warming encompasses concepts in a wide variety of fields, including fundamental ideas about how energy is transformed (physics), how chemical compounds change (chemistry), how living things interact with their environment (biology), and how carbon is cycled through the atmosphere, land, and living systems (earth science, biology, and chemistry). From this perspective, the topic of global warming is a way of tying together concepts in various fields of science, and illustrating that science is a unified way of viewing the world and identifying and solving real-world problems.

The study of global warming is an excellent way to address the goal of science for *all* students, as expressed in the *National Science Education Standards*, published by the National Research Council in 1996. We do not expect or desire all students to become scientists or mathematicians; but we *do* expect them to be sufficiently knowledgeable and motivated to remain current on new scientific research throughout their lives, and to apply what they learn to personal decisions and societal issues. Today's students are tomorrow's consumers, voters, politicians, and world leaders. The issue of global warming lies at the interface of science and societal issues, and is an excellent opportunity for students to engage in public debate now, so you can help them establish lifelong patterns of scientifically literate citizens.

Introduction

In 1990, when this unit was developed, the question of whether or not the Earth was warming due to human activities was just entering the mainstream of scientific research. Since then, the question of whether or not global warming will eventually occur has largely been resolved, and the scientific questions have shifted to determining the rate of change and what it will mean to life on our planet. In all likelihood, global warming will receive even more attention in the coming years, as the worldwide population continues to grow, and countries around the world build more power plants, manufacture more cars, and destroy more rain forests. Our students will live their entire lives during this period of global stress.

At the end of 1995, the International Panel on Climate Change, a United Nations body of 2,500 scientists from more than 100 countries, predicted that by the end of the next century, the average global temperature will climb from 3° to 8° Fahrenheit. The panel also predicted serious consequences of global warming, including rising sea level, extreme storms, droughts and heat waves, and serious impacts on human health and wildlife diversity. As a result of new scientific findings, the background for teachers (Behind the Scenes) is substantially changed and updated.

On the other hand, the student activities in this guide are changed very little. The fundamental theory of global warming and the greenhouse effect has not only survived the scrutiny of thousands of scientists; it has been elaborated and strengthened during the intervening years. There are also many scientific questions that are still unresolved, such as the effect of increased cloud cover, and the potential for the oceans to absorb excess carbon dioxide from the atmosphere.

As educators, we have a responsibility to inform our students about this topic, since it will have a high impact on their lives. The study of global climate change shows the relevance of scientific knowledge to real-life problems, and reveals the interconnections among the scientific and technical fields, as well as between science, society, and the environment—locally, nationally, and globally. There is an even more important reason for educating young people about global warming. They can do something about it. They can begin today by conserving energy. As responsible citizens, they have opportunities to influence the sources of energy that our nation relies upon; to make their voices heard about issues such as deforestation; and in some cases to take on leadership roles in the scientific, social, political, and international realms where significant questions about climate change will eventually be resolved.

This guide is designed to help you communicate the basics about global warming and the greenhouse effect to your students. Through a variety of laboratory activities, simulations, and discussions, your students will learn answers to questions such as: What is the greenhouse effect? In what way does the Earth's atmosphere act like a greenhouse? Who and what is causing the problem? What are the uncertainties of this theory? Has global warming been observed yet? How much is the Earth likely to warm up? What are the consequences likely to be? And, if we decide that global warming is something to worry about, what can we do about it?

During the unit, students improve their understanding of a number of important scientific concepts, such as the molecular model of heat, ways in which energy is transferred, how objects attain a stable temperature, and the structure of the atmosphere. They also acquire a variety of important scientific investigative skills involved in conducting and interpreting controlled experiments.

Many teachers and educational researchers find that presenting a variety of learning formats, connected to everyday social concerns, is also an excellent way to value and bridge the varying backgrounds and experiences of boys and girls, and underrepresented groups, in the classroom. Although the activities are aimed at students in grades 7–10, they can be adapted for advanced sixth grade students. The unit can also form the basis of a more extensive unit for a high school science or social studies course.

In **Session 1**, the students discuss what they have already heard about the greenhouse effect. This provides an opportunity for you to find out what your students already know, and what misconceptions they may have. The session ends with a discussion of two graphs that show global temperature records for the past 130 years, and the climatic history of our planet for the past 450,000 years.

In **Session 2**, the students perform the classic "greenhouse" experiment in which they compare the heating of an open container with the heating of a closed container. This experiment teaches the concepts of heating, cooling, and equilibrium. Students discover that a closed container of air is able to trap heat more effectively than an open container.

In **Session 3** the students play a simulation game that approximates the greenhouse effect as it occurs in the Earth's atmosphere. They observe that photons of light energy from the Sun are absorbed by the Earth and emitted as infrared (heat) photons. These infrared photons are then absorbed by carbon dioxide in the Earth's atmosphere. The more carbon dioxide, the more heat is trapped. Thus, carbon dioxide is called a "greenhouse gas." The session concludes with a discussion of several different greenhouse gases.

In **Sessions 4 and 5** the students perform a series of experiments in which they compare the relative concentration of carbon dioxide in gas samples from four different sources: ambient air, human breath, car exhaust, and a chemical reaction between vinegar and baking soda. As a homework assignment, the students find out how much of the carbon dioxide in the atmosphere is contributed by various nations.

In **Session 6** the students consider the social and ecological consequences if the Earth's climate should increase by 5 degrees Fahrenheit, including both positive and negative effects. In **Sessions 7 and 8**, they brainstorm solutions and participate in a "world conference" to decide what can be done to avoid global warming, or to reduce its detrimental effects if climate change turns out to be unavoidable.

Teachers who are unfamiliar with the background regarding global warming and the greenhouse effect may wish to read an overview of the topic in the "Behind the Scenes" section on page 125. This overview is meant for the teacher, and is not intended to be presented as a reading assignment for the students.

We use the Fahrenheit scale in this guide rather than the Celsius scale, which is widely accepted as the standard for science and technology, because Fahrenheit is the most meaningful temperature scale for people in the United States, just as miles per hour is the most meaningful scale for speed. If you want your students to use the Celsius scale, the units can be converted with the standard formulas:
°C= (°F - 32) / 1.8 and °F = (°C x 1.8) + 32.

Throughout the unit, students receive graphs of actual data to interpret and discuss. Since actual data are rarely unambiguous, many students find it quite challenging to interpret these graphs. These experiences provide opportunities for students to improve their graph-reading skills, and give them insight into why scientists often disagree with one another. Students learn that science is not simply a list of clear-cut facts to be memorized, but rather an approach to learning about the world that requires honesty, judgment, creativity, as well as knowledge and skills. Above all, they learn that science is a human endeavor that can enable people to gain an increased understanding of, and some foresight and control over, their environment.

Time Frame

The eight session outline suggested below is a minimum time frame for presenting basic information about global warming and the greenhouse effect. Many ideas are included in the text that will enable you to considerably expand this course.

Session 1: What Have You Heard About the Greenhouse Effect?

Teacher Preparation .... 30 minutes
Classroom Activity .... 40–50 minutes

Session 2: Modeling the Greenhouse Effect

Teacher Preparation .... 2 hours
Classroom Activity .... 45–60 minutes
(An additional 20–30 minutes may be needed by some classes.)

Session 3: The Global Warming Game

Teacher Preparation .... 40 minutes
Classroom Activity .... 80–120 minutes
(This session is likely to take two class periods.)

Session 4: Detecting Carbon Dioxide

Teacher Preparation .... 2 hours for Sessions 4 and 5
Classroom Activity .... 45–60 minutes

Session 5: Sources of Carbon Dioxide in the Atmosphere

Teacher Preparation .... 15 minutes
Classroom Activity .... 45–60 minutes

Session 6: Changes On Noua's Island

Teacher Preparation .... 20 minutes
Classroom Activity .... 40–50 minutes

Session 7: Worldwide Effects of Climate Change

Teacher Preparation .... 20 minutes
Classroom Activity .... 40–50 minutes

Session 8: World Conference on Global Warming

Teacher Preparation .... 20 minutes
Classroom Activity .... 40–50 minutes

What we have heard about Global Warming
Ozone and can't get out.
Suns rays enter GH
Fumes go to Ozone Layer
and cause the world to heat
Melted ice

Session 1: What Have You Heard About The Greenhouse Effect?

Overview

In this session, your students are encouraged to recall whatever they have heard about the greenhouse effect and global warming, and to think of questions they have about it. Not only is this is the usual starting place for investigations in science, but it will provide you with a way of finding out what your students already know, as well as revealing common misconceptions you will need to address during the unit.

After the students write down what they have heard, they meet in small groups to compare ideas. Then, as a class, they compile a master list of statements and questions. This list becomes a useful tool by which to assess the students' improved understanding as the unit progresses. In the last part of this session, your students participate in the ongoing scientific debate about whether or not global warming has begun by discussing graphs that show evidence of how the Earth's temperature has changed in the recent and distant past. This activity requires certain skills in reading graphs. A series of questions are suggested to help your students interpret the graphs and understand why scientists disagree about how to interpret them.

The objectives of the session are to:

(1) encourage students to begin investigations in science based on their own questions and understandings;

(2) establish a base of common information that will be useful both for teaching and evaluating the unit; and

(3) introduce the scientific debate about whether or not the average global temperature is rising.

What You Need

For the class

- ❑ 3 large sheets of butcher paper (each 2–3 feet wide and about 6 feet long)
- ❑ 1 marking pen
- ❑ 1 roll of masking tape
- ❑ 1 pair of scissors

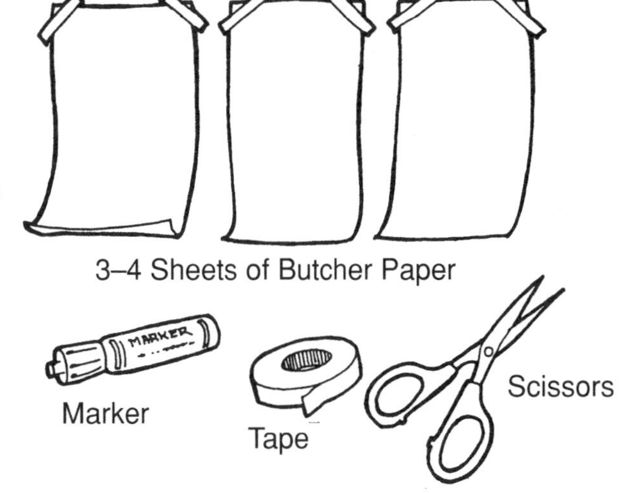

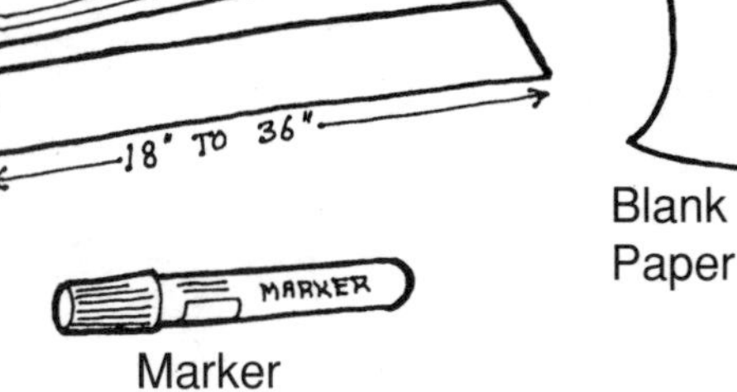

For each group of 4 students

- ❏ 1 sheet of blank paper
- ❏ 5 or more sentence strips
  (Sentence strips are commercially available from school supplies companies—see Resources on page 146—or, you can make your own by cutting sheets of butcher paper into strips about 3" wide and from 18" to 36" long—the exact size is unimportant.)
- ❏ 1 marking pen

For each student

- ❏ 1 sheet of blank paper
- ❏ 1 handout: "Average Global Temperatures . . ." (see page 15)
- ❏ 1 homework sheet: "Everyone Likes to Talk About the Weather" (see page 14)

Getting Ready

Before the Activity

1. To increase your students' interest and motivation, you may wish to announce that the topic of "the greenhouse effect and global warming" will start in several weeks, and that you want them to bring to class any newspaper or magazine articles they come across on the topic, so these can be posted on the bulletin board. However, we suggest that you do not make this activity into a formal assignment—you may drastically reduce the students' enthusiasm, rather than increase it.

You may wish to display a large sign that says:

"COMING SOON:
THE GREENHOUSE EFFECT
AND GLOBAL WARMING."

Underneath the sign, you may wish to post articles on global warming and the greenhouse effect.

2. Make copies of the handout and homework sheet—one for each student.

On the Day of the Activity

1. Find a space on one wall where you can post the students' statements and questions. Tape several sheets of butcher paper to the wall to prepare for this part of the class. If possible, leave the sheets in place after the class so the lists can be referred to in future sessions. If this is not possible, store the lists so they can be posted again later.

2. Set aside the sentence strips and masking tape for use during the class.

3. Arrange the room by pushing desks together or moving tables so there is one flat work area for each group of four students.

4. In nearly all the classes where this unit was tested, the students had heard about "the greenhouse effect," and were able to list ideas and questions. However, if your students have not heard about it, be prepared to go directly to the homework assignment "Everyone Likes to Talk About the Weather . . ." in which they ask their parents and grandparents what they know. In that event, postpone the writing and sharing activities of Session 1 until the next day.

Writing About the Greenhouse Effect

1. Introduce the topic of the greenhouse effect and global warming by telling the group you want to find out what they've heard about the greenhouse effect and how it may affect the average temperature of the entire Earth in years to come. (We've found that most students have heard of the "greenhouse effect," but not of "global warming.")

2. Emphasize that this is the same way scientists start out an investigation, by first finding out what is already known, in order to figure out what to investigate.

3. Invite the students to think about everything they've heard about the greenhouse effect—from books, TV, newspapers, or something someone said to them. It doesn't have to be long or complicated. Anything they've heard is acceptable.

4. Tell the students they have three minutes to make a list of everything that comes to mind. Tell them not to worry about using whole sentences, or whether or not they are certain of the information. Ask them if they have any questions about what they are supposed to do. Hand out pieces of blank paper, and have them begin writing.

5. If some students prefer to draw a picture of what they have heard, let them do so. However, encourage them to write a few words of summary underneath the picture. Circulate among the students, assisting those who have difficulty expressing themselves in writing.

6. If some students run out of things to write (or if they don't have any statements to write in the first place) tell them to turn over the paper and write down any *questions* they have about the greenhouse effect.

The "Mind-Swap"

1. After three minutes (or after most of the students have finished) have the students put their pencils down.

2. Explain that it is time for a "mind-swap." They will work in teams of three or four to combine what they have heard, and to find out if anyone in the team knows the answers to any of their questions.

3. Help organize the students into teams of three or four, and explain the rules as follows:

   a. In each group, one student begins by sharing all she has heard about the greenhouse effect. No one can interrupt the speaker or discuss anything yet. The other members of the group listen and, if they wish, write down the things the speaker says on their own lists.

   b. After one person finishes sharing, the next person has a turn, stating only those things on their list not covered by the first speaker. This continues until everyone has shared what was on their list.

   c. After all group members have shared information, they can ask each other questions and discuss what each had to say. Any disagreements can be discussed later.

   d. Ask if there are any questions about what the groups will be doing, and then have the groups begin.

Discussing What We've Heard

1. After about five or six minutes, stop the small group discussions. Call on each group in turn, ask for one thing that each group has heard about the greenhouse effect. List these on a large sheet of paper, so they can be referred to as the unit develops. Continue until all of the students' ideas are listed. Tell the groups it is OK to "pass" if all of their group's statements have already been listed.

2. You may want to ask if there are two people in the same group who came up with statements that contradict each other. Encourage discussion. Point out that there is often controversy and disagreement in science. You may want to explain that sometimes we think we know something, and that it might have been true at one time, but as knowledge in the scientific community grows, it is shown to be inaccurate, or is interpreted differently.

3. Point out that what each of the groups has been sharing is what the people in each group have heard to be true. Explain that as the class continues its study of the greenhouse effect, they will be able to better determine whether or not what we think we know, and what we have heard, is in fact accurate, when compared with current scientific thinking.

4. If the students have listed the "hole" in the ozone layer as a cause of the greenhouse effect, explain that many people confuse these two different environmental problems. The ozone layer refers to a concentration of ozone—a form of oxygen—about 18–20 miles above the Earth's surface. The "hole" refers to a reduction in the amount of ozone—by as much as 60% above Antarctica and 10% above North America. This means more ultraviolet light is coming to the Earth's surface, where it may cause a higher incidence of skin cancer and other diseases. While that is a very serious problem, the additional ultraviolet light does not heat up the Earth very much. The main cause of the greenhouse effect will be described in Sessions 2 and 3.

What We Don't Know About the Greenhouse Effect

1. Ask each group to discuss questions they have about the greenhouse effect, and to come up with at least one question per person. They will write each question on a strip of paper they will later post on the wall.

2. Explain the rule that a question can only be written on the strip if no one in the group knows the answer. Each group will need to discuss each question, and see if anyone knows the answer.

You may want to sort questions about the greenhouse effect into groups or categories, such as technical questions, questions about its effects, questions about the future, and questions about possible solutions.

3. Ask if there are any questions about what they will be doing. Distribute strips of paper and a marking pen to each group, and have them begin.

4. After about five minutes of group discussion, regain the attention of the class and have each group post the questions they generated.

5. After all groups have posted their questions, ask if anyone in the class can offer an answer to any of them. Tell the class you will return to this list several times during the unit, to see which questions have been answered.

6. Point out that scientists and others studying the greenhouse effect do not know all of the answers to the questions about it. That is one reason why it is such an interesting subject to study; we are learning right along with the scientists!

The following is an example of a list of student statements and questions about the greenhouse effect, generated by a 9th grade class.

Students' statements about the greenhouse effect

- The sun's rays all come into the greenhouse, and because of insulation in the greenhouse, [they] can't escape.
- People are worried that the Earth's going to get so hot that everyone on Earth is going to die.
- People are thinking about the drought. The reason we might be having one [in California] is [because] there is a wall between the hot air and the cold air, so it doesn't circulate.
- Heating will melt the ice caps and the seas will go up.
- There are so many pollutants in the atmosphere that we can't really tell what's causing it.
- I hear that all the water is gonna dry up and like everything is gonna die and everything, and the Earth is gonna become like the moon.

Students' questions about the greenhouse effect

- Will the seas dry up, and if so, when?
- Is it possible for animals to adapt to the greenhouse effect?
- Will there be less rain because of the water evaporating?
- Can we get cancer?
- Can we patch the ozone?
- Will there be an Earth in the year 3050?
- Will we all have to live in [glass or plastic] bubbles?

Thinking About Climate Change

If your class is confused about the difference between climate and weather, this is a good time to clarify the difference. Weather is the condition of the atmosphere at a particular time and place. Climate refers to the average weather conditions, determined over a period of about 30 years, for an entire region or the whole planet.

1. Ask the class what evidence they think there is for a change in the Earth's climate. Have they heard any comments about changes in the climate where they live?

2. Point out that some scientists now think the average temperature of the Earth is increasing. Ask the class how they would measure a change in the overall temperature of the Earth.
   [The temperature would need to be measured at many different locations several times each day, and all of these numbers would be averaged each year.]

3. Hand out copies of "Average Global Temperatures in the Recent and Distant Past" (see page 15). Direct attention to the upper graph, and ask a volunteer to read what the scales represent.
   [Horizontal axis = years from 1860 to 1995;
   vertical axis = average global temperature in °F]

4. Ask the class what has happened to the average temperature of the Earth over this period.
   [It has gone up and down a lot, but there has been a consistent rise.]

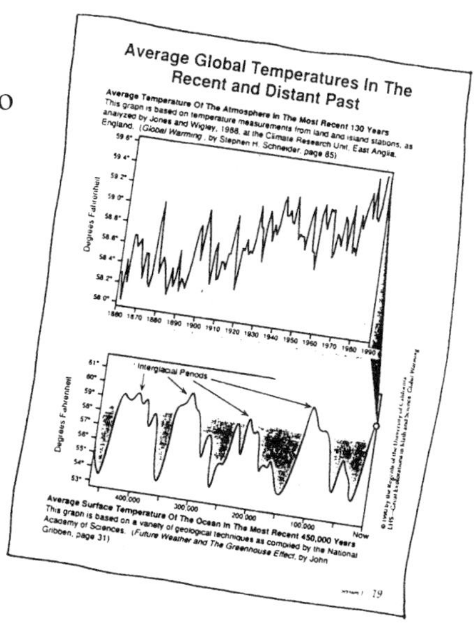

5. Point out that scientists are concerned about this increase in temperature, not just because it is going up steadily, but because we are currently in one of the warmest periods of the Earth's history. That history is shown in the lower graph.

Scientists determined how the temperature of the earth changed over thousands of years by studying the composition of polar ice. Each year's snowfall forms a separate layer, like layers on a cake. Scientists have been able to determine the average temperature of the Earth by analyzing the ice within each layer. (See page 134 for a more detailed explanation.)

6. Ask a volunteer to read what the scales in the lower graph represent.
[Horizontal axis = time before the present;
vertical axis = surface temperature of the ocean in °F]

7. Ask the class to determine how much the temperature of the Earth has changed over the past 450,000 years.
[It has varied from about 54°F to 60°F.]
Explain to the students that this 6°F change was in the **surface temperature of the ocean**. Scientists think the average air temperature of the Earth followed the same pattern, but changed by about 9°F or 10°F. (This is because the polar ice caps insulate the ocean so its temperature does not change as much as the atmosphere.)

8. Tell the class that when the average temperature of the Earth's atmosphere cools by only 9°F or 10°F, the Earth experiences a major Ice Age. During this time, there is much more ice and snow on mountains, and the north and south polar caps extend much further towards the equator. Ask: "How long ago was the last warm period, such as the one we are in today?"
[About 100,000 years ago]
Point out that a warm period is called an "interglacial," which means time between Ice Ages.

9. Now ask the students to meet in small groups and discuss the two graphs, and answer the following series of questions (write them on the board):

   - During the past 450,000 years, would you say that the Earth has been generally warmer than now, or colder than now?
     [Colder]

   - How many warm periods, or "interglacials," have there been in the past 450,000 years?
     [Five, including the present]

- About how long has the Earth been in its current "warm period"?
  [About the last 12,000 years]

- How much has the atmospheric temperature changed in the past 110 years, since 1880?
  [From 58°F to 59°F, a rise of about 1°F]

10. Ten minutes before the end of the session, ask the groups to report their answers to the questions. Discuss and resolve disagreements.

11. Point out that some scientists predict the Earth will cool off, because of the cyclic pattern of temperature change in the past. Others say human activities are changing the pattern. They predict that, if we do not change the way we use energy, the average global atmospheric temperature is likely to increase an additional 2°F to 9°F by the middle of the next century. They predict the Earth will become hotter than it has ever been in the past 450,000 years.

Introducing the Homework

1. Hand out the homework assignment sheet "Everyone Likes To Talk About The Weather . . .".

2. Ask the students to interview older people who may remember what the climate was like 50 or 60 years ago, and to find out whether they think the climate has changed. The students might interview grandparents, other older relatives, or neighbors; and discuss the questions on the homework sheet.

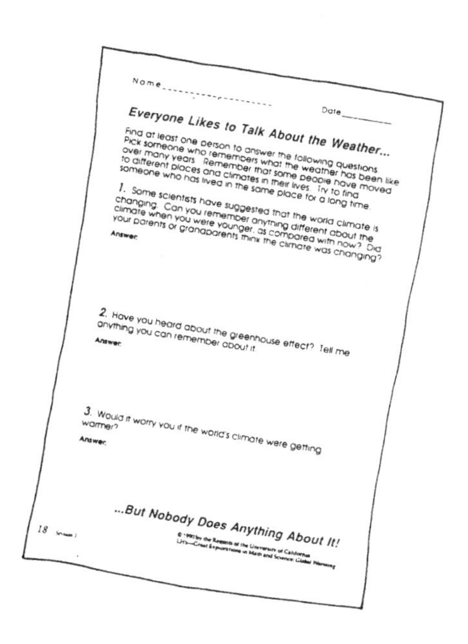

Name ____________ Date ________

Everyone Likes to Talk About the Weather...

Find at least one person to answer the following questions. Pick someone who remembers what the weather has been like over many years. Remember that some people have moved to different places and climates in their lives. Try to find someone who has lived in the same place for a long time.

1. Some scientists have suggested that the world climate is changing. Can you remember anything different about the climate when you were younger, as compared with now? Did your parents or grandparents think the climate was changing?

Answer:

2. Have you heard about the greenhouse effect? Tell me anything you can remember about it.

Answer:

3. Would it worry you if the world's climate were getting warmer?

Answer:

...But Nobody Does Anything About It!

18

Name_______________________________________ Date _________________

Everyone Likes to Talk About the Weather . . .

Find at least one person to answer the following questions. Pick someone who remembers what the weather has been like over many years. Remember that some people have moved to different places and climates in their lives. Try to find someone who has lived in the same place for a long time.

1. Some scientists say the world climate is changing. Can you remember anything different about the climate when you were younger, compared with now? Did your parents or grandparents think the climate was changing?

Answer

2. Have you heard about the greenhouse effect?
Tell me anything you know about it.

Answer

3. Would it worry you if the world's climate were getting warmer?

Answer

. . . But Nobody Does Anything About It!

Average Global Temperatures in the Recent and Distant Past

Average Temperature of the Atmosphere in the Most Recent 130 Years

Based on temperature measurements from land and island stations, as analyzed by Jones and Wigley, 1988, at the Climate Research Unit, East Anglia, England. (*Global Warming* by Stephen H. Schneider, page 85)

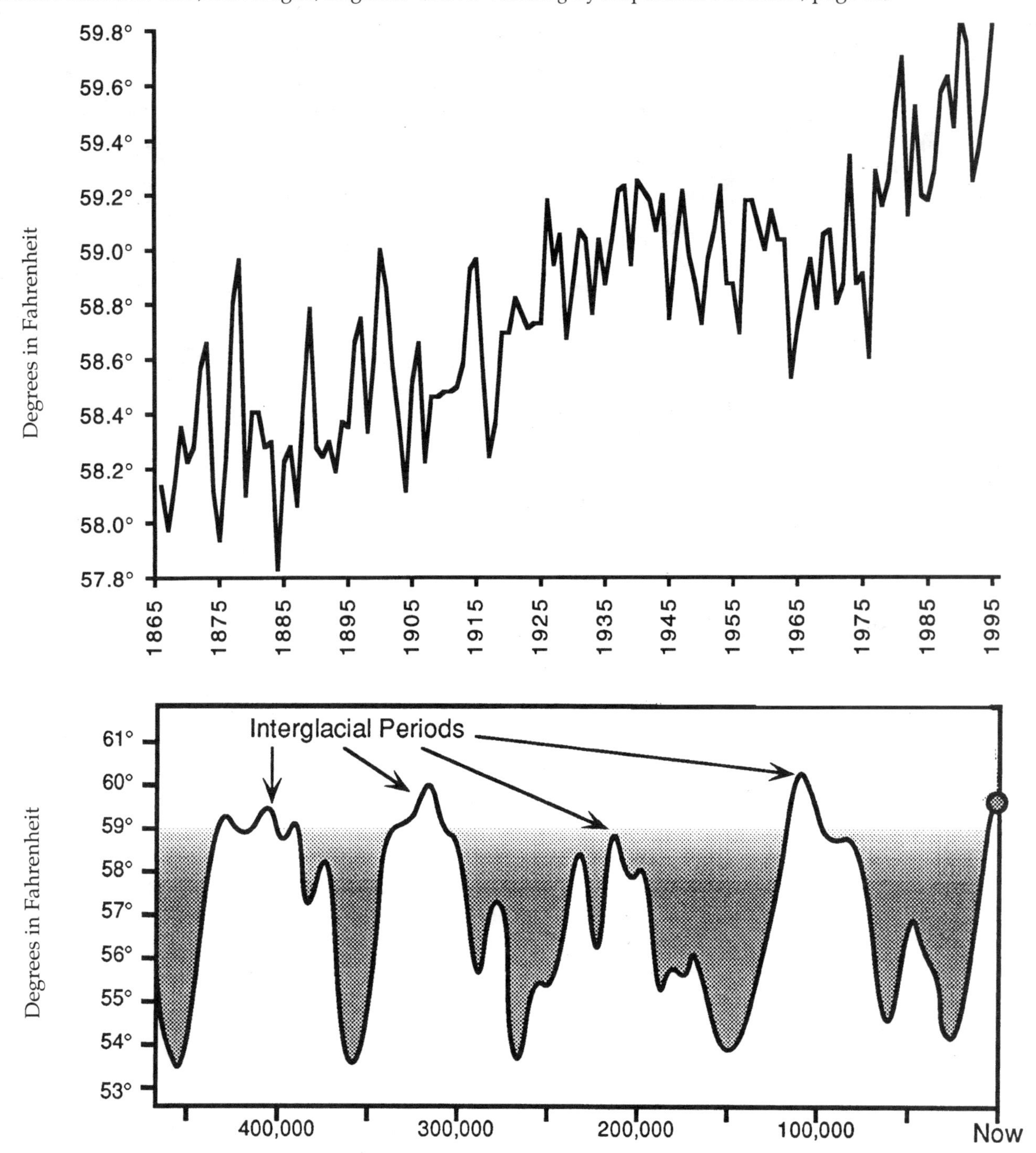

Average Surface Temperature of the Ocean in the Most Recent 450,000 Years

Based on a variety of geological techniques as compiled by the National Academy of Sciences. (*Future Weather and The Greenhouse Effect* by John Gribben, page 31)

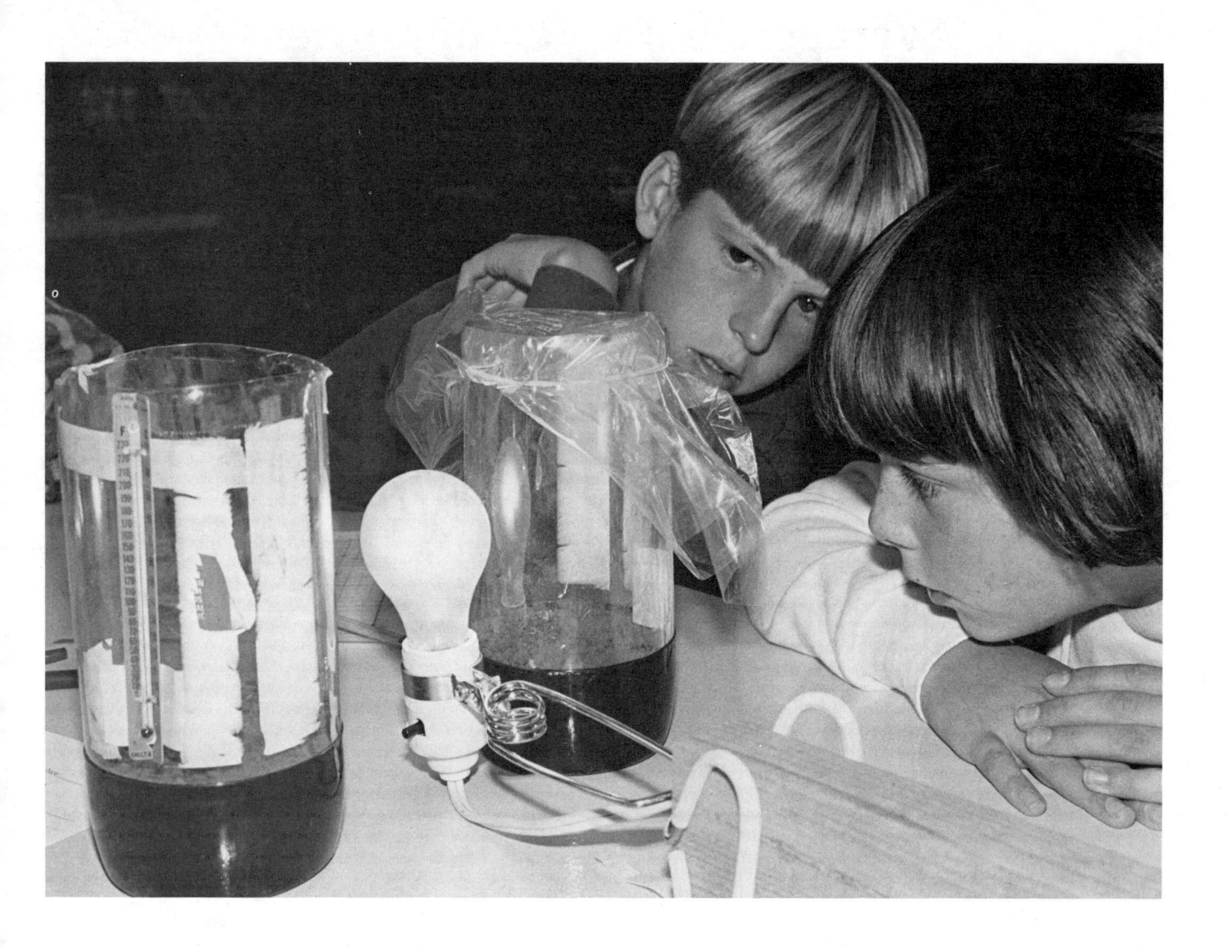

Session 2: Modeling the Greenhouse Effect

Overview

In this session, the students perform an experiment to learn about the greenhouse effect. By constructing a physical model of the atmosphere using familiar materials, the students discover that air trapped in a container will heat up more than air in an open container, when both are exposed to the same amount of energy from a light bulb. In a comparable (but different) way, the carbon dioxide in the atmosphere acts as a "heat trap" for energy from the Sun.

The objectives of this session are to: (1) provide basic information about the Earth's atmosphere and how scientists use models to study it; (2) provide students with an opportunity to build and test a physical model analogous to the atmospheric greenhouse effect; (3) introduce the concepts of energy transfer and thermal equilibrium; and (4) give students practice in setting up controlled experiments, accurately recording data, graphing, and interpreting the results.

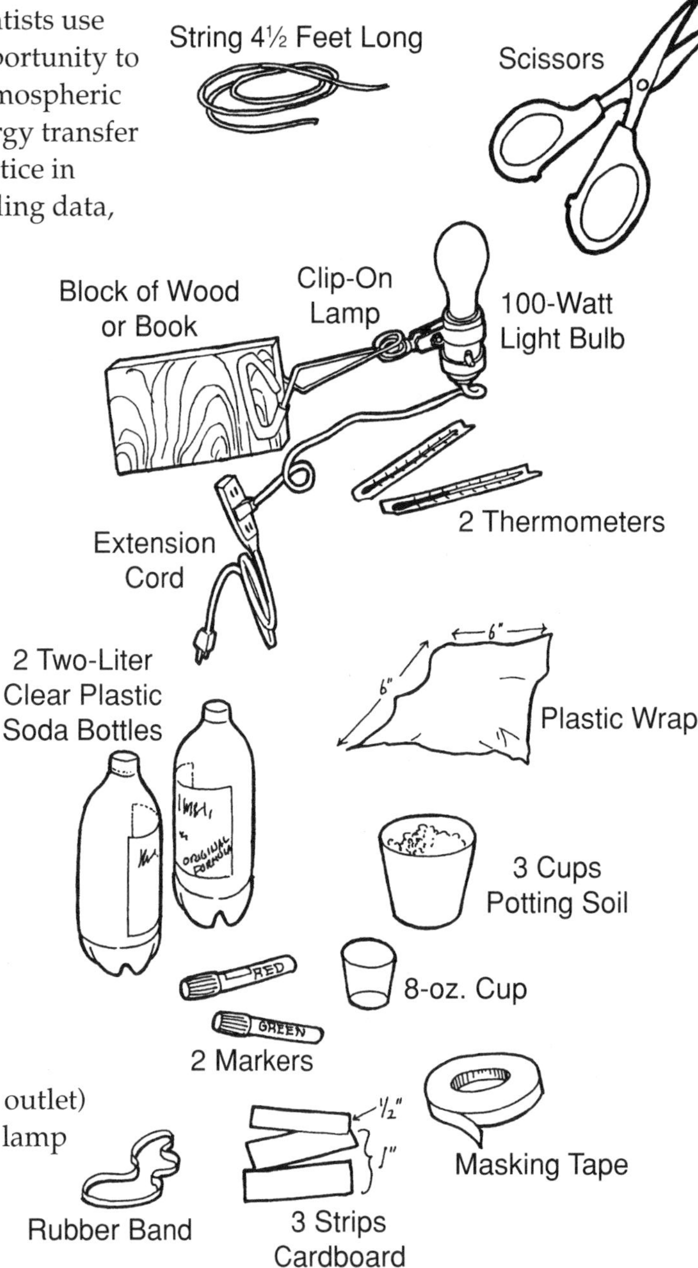

What You Need

For the class

- ❑ 1 piece of string, 4 1/2 feet long
- ❑ white chalk and chalkboard, or marking pen, and butcher paper
- ❑ 1 pair of scissors

For each group of 4 students

- ❑ 2 two-liter clear plastic soda bottles
- ❑ 2 thermometers
- ❑ 3 strips of thin cardboard, 1/2" x 1"
- ❑ 3 cups of potting soil
- ❑ 2 fine or medium-tipped marker pens (one red, one green)
- ❑ 1 roll masking tape
- ❑ 1 piece plastic wrap, approximately 6" x 6"
- ❑ 1 rubber band
- ❑ 1 100-watt light bulb
- ❑ 1 clip-on lamp
- ❑ 1 extension cord (if needed to connect lamp to outlet)
- ❑ 1 large book or piece of wood for securing the lamp
- ❑ 1 plastic 8-oz. cup
- ❑ 2 graphing data sheets: "The Greenhouse Effect" (see page 28).

Getting Ready

Before the Day of the Activity

1. Collect 2-liter plastic soda bottles (two for each team of four students). Cut off the top at the point where they begin to narrow at the neck.

2. Tagboard or a manila folder can be used to make a strip of thin cardboard, 1/2" wide by 1" long, to be used as a spacer to place the two bottles exactly half an inch from the light bulb. Cut two squares of cardboard to cover the bulbs of the thermometers so they are shaded from the direct rays of the light bulb.

On the Day of the Activity

If you have extra time to devote to this session, allow your students to prepare the experimental equipment themselves. Alternatively, you may want some student volunteers to help you set up the equipment before class.

1. Tie the chalk onto one end of the string. Standing next to the board, place your foot on the free end of the string and practice drawing an arc with a 4-foot radius (see page 21).

2. Make copies of the graphing data sheet "The Greenhouse Effect." Give every two students one copy (see page 28).

3. Close windows, window shades, and doors so the classroom is free of drafts and direct sunlight.

4. Gather the clip-on lights, light bulbs, and stands. Check that there are enough extension cords to reach the available outlets. Each bulb should be placed so it is standing upright between two bottles.

5. Tape the thermometers and cardboard strips inside the bottles as shown in the drawing on page 19. If the bottles still have labels attached, tape the thermometer just to one side of the label, on the inside of the bottle, so the label doesn't interfere with the experiment.

6. Using a cup to measure, put about 1½ cups of dry potting soil into each bottle.

7. Have one set of equipment near the front of the room to demonstrate how to set up the experiment.

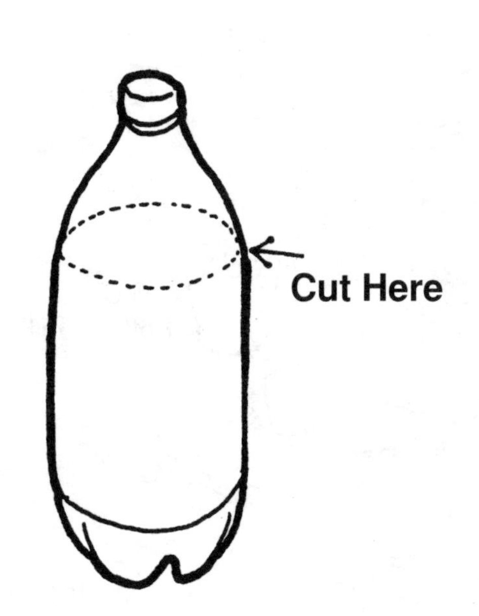

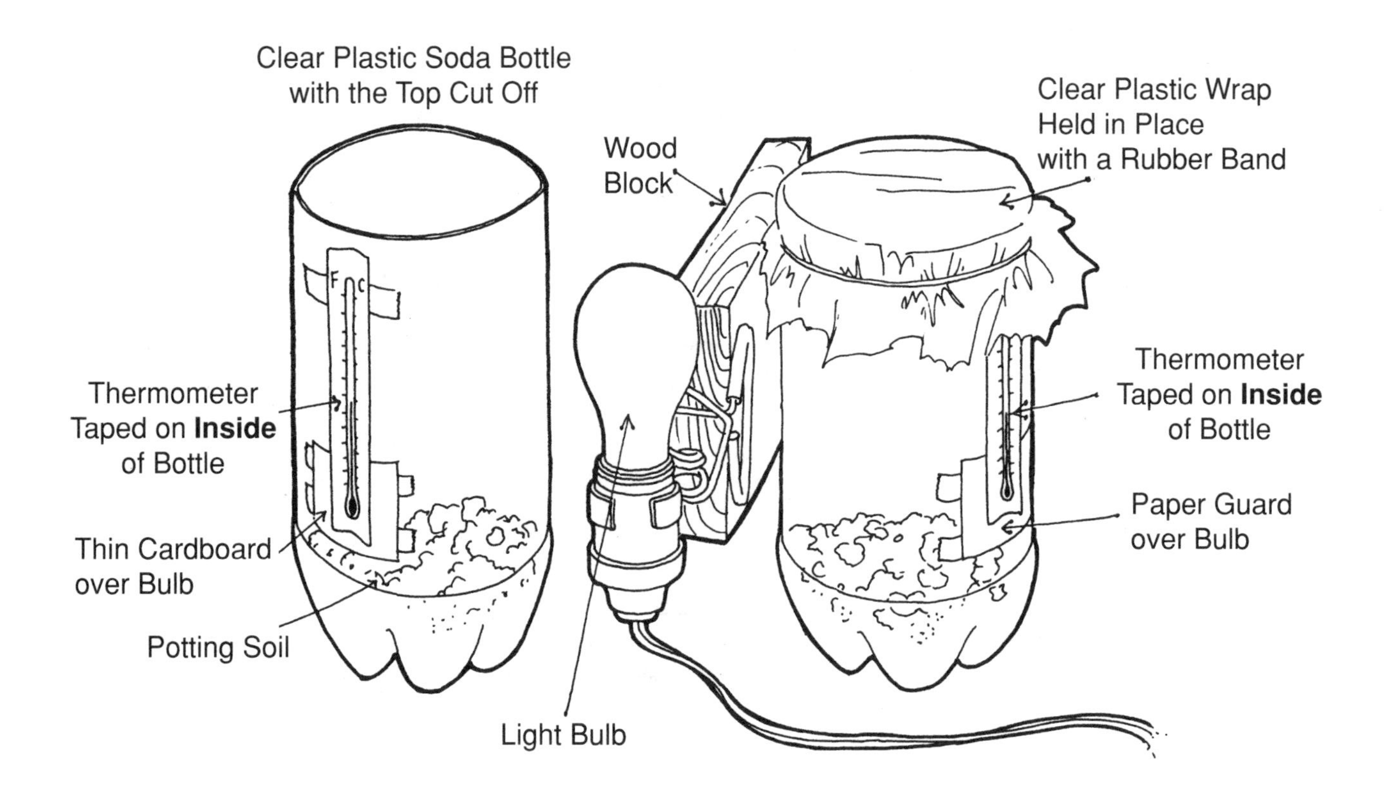

This is how each group's setup should look at the start of the Greenhouse Experiment.

Discuss The Homework

1. Invite the students to discuss their answers to each of the questions on the homework sheet. Emphasize that it is very difficult to determine whether the climate is changing, because of the variability in weather patterns from year to year.

2. Point out that some ***climatologists*** (scientists who study climates) think the climates in the world might be starting to change in a major way, due to human activity. They believe the data the class looked at in the first session supports this idea.

3. Explain that one of the purposes of this unit is to give students background knowledge about the ***greenhouse effect,*** which some climatologists think may be causing a warming up of the Earth.

Why Do We Need a Model of the Atmosphere?

1. Point out that the atmosphere is a large and complex system, so experiments and measurements concerning it are difficult to perform.

2. Ask the class to suggest experiments or ways of measuring the average temperature of the Earth's atmosphere. [Averaging lots of temperature measurements; finding places with long histories of records; analyzing data from isolated places, such as islands, which are less affected by other changes that can affect the climate.]

3. Ask what difficulties scientists might have in determining whether or not the average temperature of the Earth is heating up.
   [It is difficult to: find long-term historical data; regularly measure the temperature at sea and at the polar ice caps; distinguish long-term changes in average temperature from short-term variations; take into account differences in temperature between seasons and between places; and find locations that are unaffected by local factors, such as urban development and deforestation.]

4. Remind students that the greenhouse effect and global warming are new areas of scientific study, and scientists do not have all of the answers, partly because of the difficulties the class has just discussed.

5. If your students have not already done so, point out that one way to test theories about climate change is to build a model of the atmosphere, and to experiment with the model. That's what the class will be doing this session.

6. Explain that before they begin to build their models, they need to have a clear idea of what the atmosphere is like.

How High Does the Atmosphere Go?

1. Tie a piece of chalk to a length of string. Standing next to the chalkboard, place your foot on the free end of the string, and draw an arc on the board, with a radius of about four feet. (See the illustration on the next page.) Tell your students that this arc represents the surface of the Earth.

Chalkboard

White Chalk Line in Arc

2. Invite the students to suggest how far the Earth's atmosphere would extend above the surface in this drawing.
   [Students will probably suggest anywhere from two or three inches up to several feet.]
   Indicate their suggestions on the board, above the chalk line.

3. Tell the students scientists have found that over 90% of the Earth's atmosphere is within about 10 miles of the Earth's surface. The distance from the center of the Earth to its surface equals about 4,000 miles, and the scale of this drawing is about 1 foot = 1,000 miles. On this scale, 10 miles is about ⅛th of an inch, about as thick as a chalk line. In other words, 90% of the Earth's atmosphere lies within the thickness of the chalk line used to draw the Earth's surface!

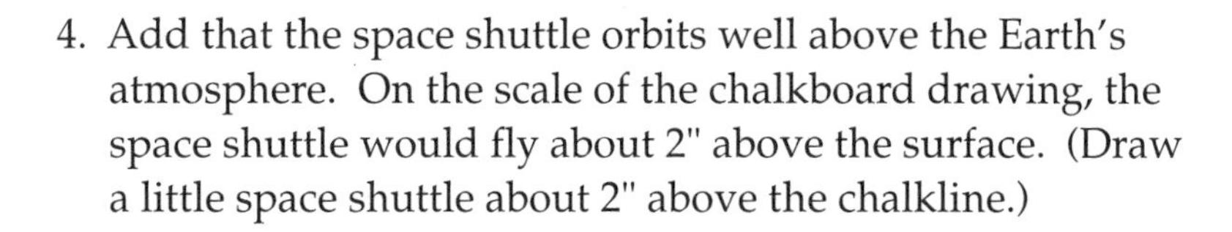

4. Add that the space shuttle orbits well above the Earth's atmosphere. On the scale of the chalkboard drawing, the space shuttle would fly about 2" above the surface. (Draw a little space shuttle about 2" above the chalkline.)

5. Explain that another way of understanding how far out the atmosphere extends is to imagine the Earth shrunk to the size of an apple. At that scale, the atmosphere is only the thickness of the skin of the apple.

A Closer Look

1. Draw a rectangle around a small part of the curved line that represents the Earth's atmosphere, and explain that you are going to magnify that part about 200 times.

2. Draw another rectangle about two feet tall, with connecting lines to the smaller rectangle, showing it as an enlargement of that area. Label the base of the large rectangle "the ground," and label the top " 10 miles high."

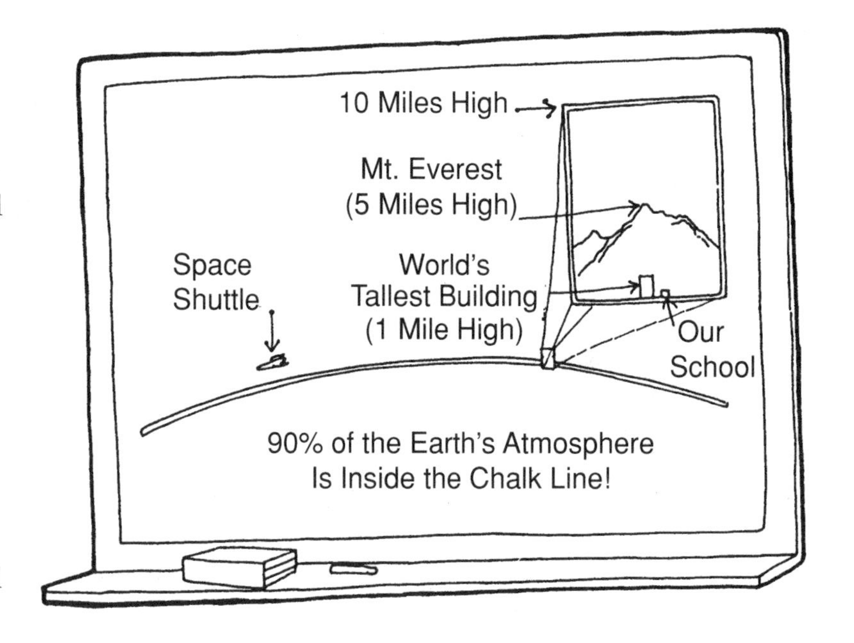

3. Ask the students, "How high is the tallest mountain in the world?"
   [Mt. Everest, over 29,000 feet, or more than 5 miles above sea level]
   Draw a representation of Mt. Everest, with a peak a little more than halfway up the rectangle. Add other drawings to the rectangle, such as the world's tallest building—about one mile—and the height of the school. Ask, "Can our school be seen on this scale of 1 mile = 2.4 inches?"

4. Point out exactly where the atmosphere ends is a debatable point. The lowest 7–10 miles of the atmosphere is called the troposphere. The stratosphere goes up to about 30 miles. Together, the troposphere and stratosphere contain 99.9% of the air. (You may want to mention that ozone is spread throughout the stratosphere, concentrated at a height of about 18–20 miles, which is called "the ozone layer." Emphasize that the "hole" in the ozone layer does not cause the greenhouse effect.)

5. Stress that most of the atmosphere is within the rectangle drawn on the board, and it gets thicker as you go down (as every mountain climber or high-flying pilot knows). Ask the class why it is that the atmosphere is thicker (denser) at the bottom—what holds it down?
   [Gravity]

6. Tell the class they are now going to make an experimental model of the atmosphere, and explore what happens when light shines on it.

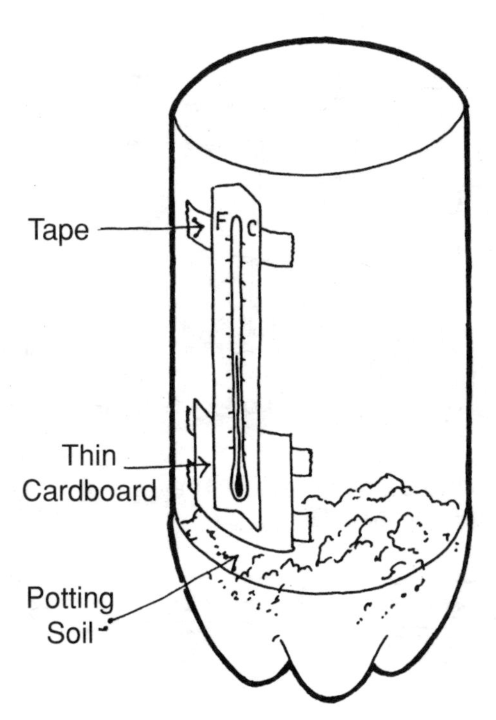

The Greenhouse Experiment

1. Assemble students into groups of four (or three). Do not distribute any equipment yet. Hold up a cut-off plastic soda bottle. Explain that the air in the bottles is going to "model" the Earth's atmosphere when it is exposed to rays from the Sun. Ask the class "What is going to model or represent the Sun?"
   [The light bulb]

2. Explain that because the experiment is going to measure the temperature of the air when exposed to the light rays, each bottle needs to have a thermometer inside. Hold up one of the bottles and point out:

a. A thermometer is taped to the inside of the bottle, above the soil.

b. The scale can be read through the plastic.

c. The thermometer bulb is covered with paper to protect it from direct rays from the light bulb

Use the Spacer to Position the Bottles

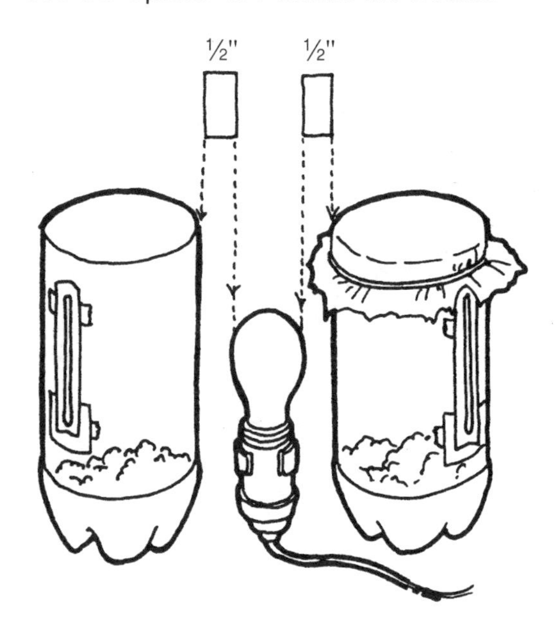

3. Explain that one of the bottles will be the ***control*** and nothing more will be done to it. Show the students how to put plastic wrap over the top of the second bottle by placing the plastic wrap tightly across the top, and securing it with a rubber band.

4. Pass out the equipment. Have each group cover one of the bottles with plastic wrap and secure it with a rubber band.

5. Demonstrate how to set up the experiment.

   a. Arrange the bottles on either side of the light with thermometers facing outward, so they can be read easily.

   b. Space the bottles equally distant from the light, using a half-inch strip of paper as a spacer. The light bulb should be turned off at this time.

6. Ask the students if they think the two thermometers should read the same or different temperatures at this time.
   [They should all show the same temperature—room temperature.]
   Have the students read the two thermometers. If they are different, explain that they must add degrees to the reading of the thermometer that gives a lower temperature, so they read the same at the start of the experiment. They will also need to add this number to each reading of the "lower" thermometer, as the experiment progresses. Tell them to write this number of degrees on a note next to that thermometer so they do not forget. This is called ***zeroing*** the thermometers.

If you notice a team that has set up an experiment near a draft or in direct sunlight, find a more sheltered area for that group's experiment.

7. Have students set up their experiments. Assist groups who are behind the others in setting up, so that all groups will be ready to start the experiment at the same time.

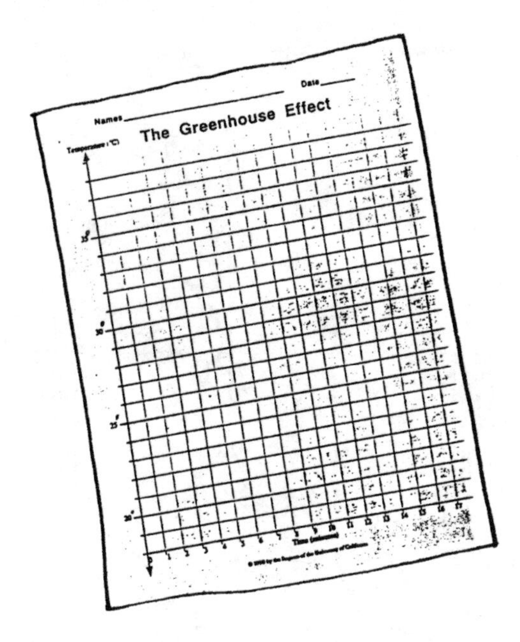

8. When all of the groups are ready to begin, ask them to listen carefully. Hold up the data sheet "The Greenhouse Effect." Explain that this is what each group will use to record their results. Have two students in charge of one bottle and two students in charge of the other. One student (the observer) will read the thermometer and one student (the recorder) will use a *pencil* to record the data directly onto the graph.

9. Ask the students to vote on which bottle they think will get hotter, and by how much, when the light is turned on.

10. Hand out the data sheets (two per group) and pencils, at the same time checking each group's setup.

11. When you are sure that everyone is ready, explain that they will take a thermometer reading and record it with a pencil mark on their graph once every minute for 15 minutes. Have the students record the temperature (at time = 0). Begin the experiment by saying "Ready, Set, Go!" as they turn on their light bulbs.

Some of your students may be disappointed when the temperatures don't suddenly go up. Reassure them that there will be changes, but the temperature change could take a few minutes to become evident.

12. Call out the time each minute, for 15 minutes, or a few minutes longer if the temperatures in all bottles have not leveled off. Circulate to make sure the groups are recording the data correctly.

Analyze the Data

1. When the experiment is finished, tell each pair of students within each group to swap data with the other pair, and to copy this other set of data, in pencil, directly onto their graphing sheets. Tell them to label the graphs so that each pair has a result from a control bottle (uncovered) and an experimental bottle (covered).

2. Hand out red and green pens. Tell the students to draw a green line between dots for the control (open) bottle, and a red line between dots for the experimental (closed) bottle.

3. Have each group write, in large letters, the names of the members of the group on one of their completed graphing sheets. Collect one sheet from each group, and arrange them on the wall or board in a way that will allow the students to compare them easily.

4. Ask one student from each group to summarize what

happened to the air temperature of the bottles in their experiment. Indicate which of the experimental results posted on the wall they are describing.

5. Ask the class to summarize what trends they see when they compare the experimental and control bottles. [In most cases students observe that in each bottle, the temperature increased, then leveled off, and at the end of the experiment, the open bottle is cooler than the closed bottle.]

If it is a very hot day, there may not be a large difference between the air temperature in the closed bottle and the open bottle. In any case, it is not essential for the experiments to come out exactly as you expect. Encourage the students to discuss what they actually observe.

6. Ask a series of questions to help the students explain the results of their experiments in their own words.

   - Why did the temperature in each of the bottles go up? [Both light and heat from the bulb passed through the plastic and warmed the air and soil inside the bottle.]

   - Why did the temperature of both bottles level off? [Heat from the light bulb can get out as well as in. To see this, turn off the light bulb and hold your hand behind the bottle. You will feel heat coming out.]

   - Why did the temperature of the closed bottles level off at a higher temperature than the open bottles? [The air inside both bottles is heated. The warm air in the open bottle mixes with cooler air outside, while the warm air in the closed bottle is trapped by the plastic top.]

Understanding the Greenhouse Effect

1. Explain to the students that the point at which temperature levels off is called the ***equilibrium temperature***. It is called that because the flow of energy into the bottle just equals the flow out of the bottle. Ask the students, "What difference did you measure between the equilibrium temperatures of the experimental and control bottles?"

2. Ask the students to tell you what happens on a hot day in a car parked in the sun with the windows shut.
   [It gets very hot.]
   Does the temperature keep going up as long as the Sun shines?
   [No, it levels off at a higher temperature.]
   How do you cool the car down?
   [By opening the windows]
   Why does that work?
   [Because the hot air can get out, and cool air can get in]

3. Explain that this phenomenon of heat being trapped, as it was inside their closed bottles, is called the ***greenhouse effect***. It is called this because greenhouse buildings, made of glass and used for growing plants, trap warm air in the same way.

4. Remind the students that the purpose of the experiment was to make a model of the atmosphere, and explore what happens when light shines on it. Ask them to relate their model to the real world, using questions such as the following:

   - How did the equipment used in this experiment correspond to the real Earth?
     [The bulb was similar to the Sun; the air represented the atmosphere; and the soil was like the Earth.]

   - How did the sample of air inside the bottles behave in a way that is similar to the Earth's atmosphere?
     [The air heated up when exposed to heat and light from the Sun, and finally leveled off at a certain equilibrium temperature.]

   - What things were different between this experiment and the real Earth's atmosphere?
     [The Earth's atmosphere does not have any solid barriers, like the plastic.]

*Some students might suggest the open bottle represents the "hole" in the ozone layer. If this comes up, ask the students what they think would happen to the temperature if you poked a hole in the plastic cover [The temperature would go down.] Emphasize that the thinning of the ozone layer is **not** a major cause of global warming.*

5. Conclude by explaining that scientists think something is trapping heat in the Earth's atmosphere, causing the temperature to go up. However, it is not a solid barrier like the plastic. The students will determine what is causing the greenhouse effect in the atmosphere in the next session.

Going Further

1. Partially cover a number of experimental bottles, rather than completely covering them with plastic at the beginning of the experiment, and then compare the final equilibrium temperature of the various experimental bottles. Classes that test this option find the results are not always predictable since minor drafts and other air disturbances in the room can have major effects.

2. A variation would be to conduct the experiment by placing a glass jar in sunlight near the window, placing the thermometers inside the glass jars.

3. For groups whose bottles have reached equilibrium, have the students punch a small hole in the plastic covering of the experimental bottle, and continue to monitor the temperature until it reaches a new equilibrium. This will emphasize the dynamic nature of the experimental results.

4. Students can compare the effects on equilibrium temperatures in covered bottles by adding water or different sorts of dried earth, as these have different absorptive and reflective capacities.

5. Students could act out the results of the experiment they conducted in Session 2 and demonstrate what happens to the air inside a closed or open bottle.

Names ______________________________ Date ____________

The Greenhouse Effect

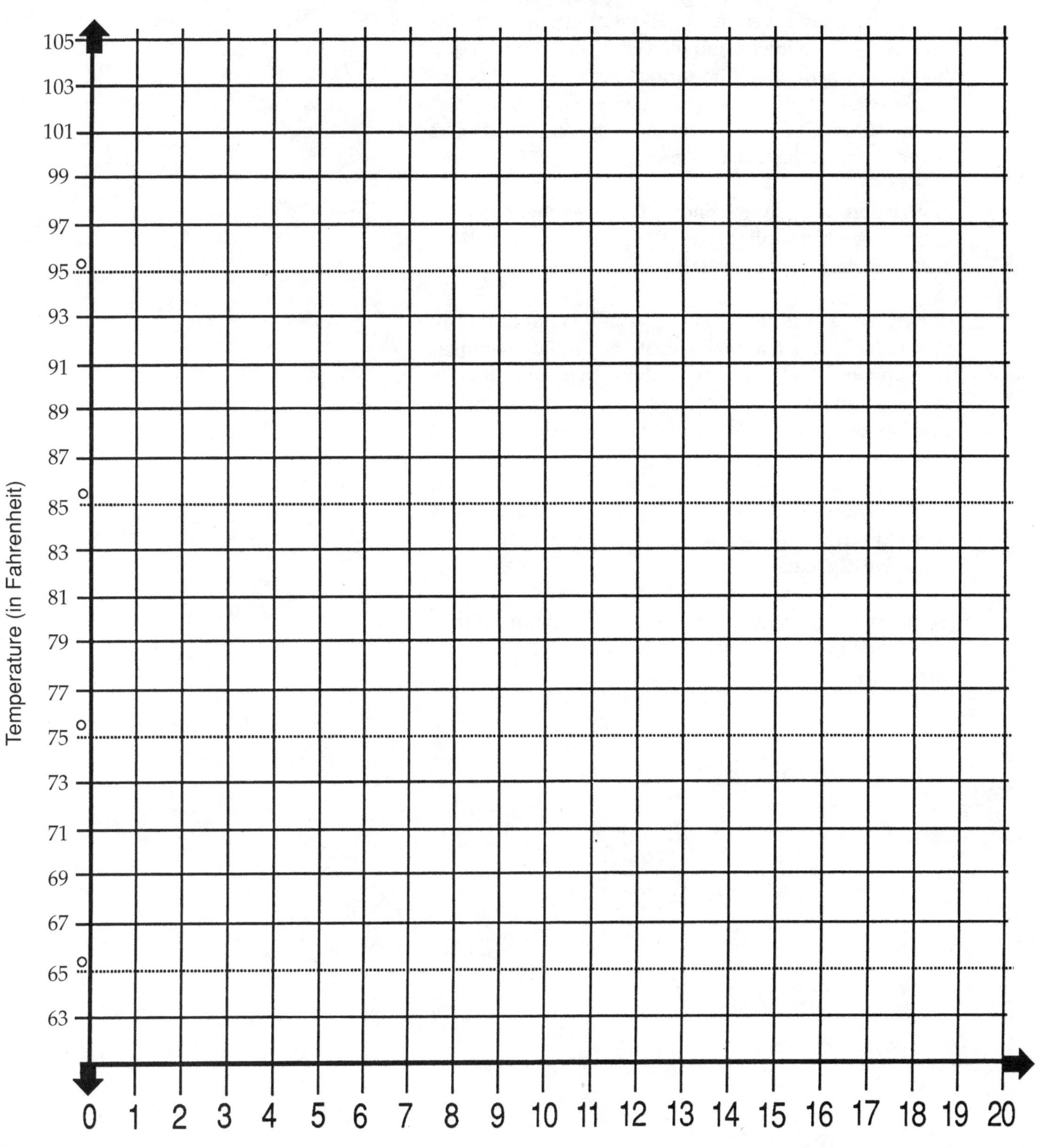

Session 3: The Global Warming Game

Overview

In the previous session, the class performed an experiment to observe the greenhouse effect. That model was a good analog of the greenhouse effect in a car or a greenhouse: however, the experiment is not a very accurate model of the greenhouse effect in the Earth's atmosphere, where there is no solid barrier, and where warm and cool air masses mix freely.

This session demonstrates the heat trap that occurs in the Earth's atmosphere, using an enjoyable board game to show what happens to the light energy from the Sun when it enters

the Earth's atmosphere. As an introduction to the game, the students learn, in a short dramatization, about the concepts of visible and infrared photons of energy, molecules, and heat. They learn that when a photon encounters a molecule its energy can be absorbed, and this causes the molecule to vibrate. We experience these molecular vibrations as heat. When a molecule cools off, it stops vibrating, and releases an infrared photon.

This session requires a longer period of time than the other sessions in this unit. Schedule about 90–120 minutes for all of the activities. If you need to divide these activities between two class periods, we suggest you introduce the game with the drama, and allow the students to play Round 1 in the first period. Have the students complete Rounds 2 and 3 and discuss their insights about global warming in the second period.

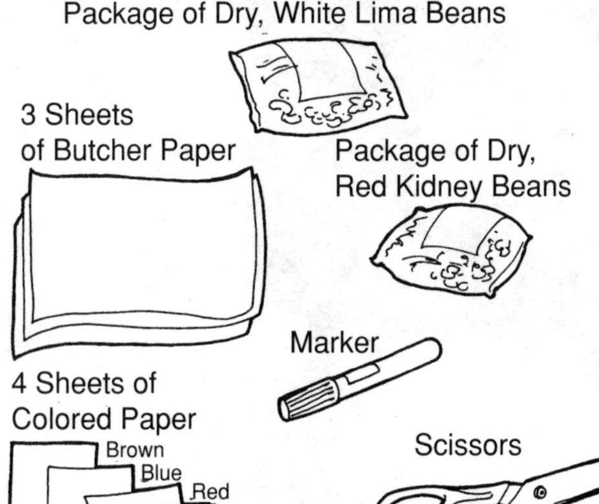

The purposes of this session are to:

(1) introduce the concepts of visible and infrared photons, showing how visible and infrared light interact with molecules in the Earth and atmosphere, and how this relates to our experience of "heat";

(2) allow students to explore the relationship between the amount of carbon dioxide in the atmosphere and the extent to which energy from the Sun is trapped in the molecules of the Earth's atmosphere; and

(3) further investigate evidence for links between the greenhouse effect and global warming.

What You Need

For the class

- ❑ 3 large sheets of butcher paper
- ❑ 1 felt-tip pen
- ❑ 4 sheets of colored paper (yellow, red, blue, and brown)
- ❑ 1 lamp and wood block stand from Session 2, for demonstration
- ❑ 1 package (about 1 lb.) of large, dry, white lima beans
- ❑ 1 package (about 1 lb.) of large, dry, red kidney beans (any large beans or other red and white objects can be substituted)
- ❑ 1 roll of transparent tape
- ❑ 1 pair of scissors

optional

- ❑ 1 calculator

For each group of 4 students

- ❑ 1 copy of the "Global Warming" game board and cards (see pages 46–54)
- ❑ 3 copies of the score sheet (see page 48)
- ❑ 1 penny or other coin

For each student

- ❑ 1 handout: "Global Temperatures and Carbon Dioxide" (see pages 56–57)

Some teachers suggested painting one side of the beans red (with spray paint for many beans, or permanent red marker for a few) then, instead of switching beans from white to red, students could just flip the bean over for it to represent an infrared photon.

Getting Ready

Before the Day of the Activity

You may want to ask for student volunteers to help you prepare some of the materials.

Some teachers made overhead transparencies of the game board, and used tinted plastic chips or bits of colored acetate to demonstrate the rules of the game.

1. Assemble the game boards for each group, plus one for demonstration, by photocopying two game sheets (pages 46–47) and joining them with tape. Photocopy and cut out the three score sheets (page 48) for each team.

2. Photocopy and cut out the three sets of game cards (pages 49–54) for each group. Copying the card sets onto paper of different colors makes them a little easier to use, but this is not essential. Suggested colors are yellow for "Heads- Reflected," red for "Tails-Absorbed," and blue for "Carbon Dioxide."

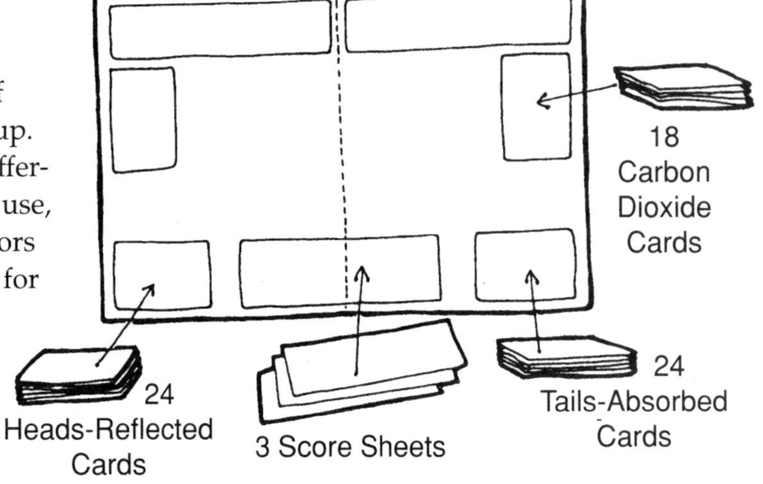

3. Using a marking pen and full-size sheets of colored paper, make signs that can be read from the back of the class:

   Yellow .... VISIBLE LIGHT PHOTON
   Red .... INFRARED PHOTON
   Blue .... CARBON DIOXIDE MOLECULE
   Brown .... ROCK MOLECULE

4. Using butcher paper and a felt-tip pen, make three large signs that say:

   OUTER SPACE
   THE ATMOSPHERE
   THE EARTH

The photon sign could be made from two sheets of paper, one red and one white. The two sheets could then be glued or taped back-to-back with "visible light photon" written on the white side, and "infrared light photon" on the red side. During the drama, the student playing the photon just flips the paper over when the photon is absorbed and re-emitted.

The game boards can be glued to a manila folder, and laminated with contact paper. The game pieces can be stored inside the folder in a small plastic bag. This whole set can then be used many times. It needs to include an erasable pen that can write on plastic, to add the CO_2 molecules in Rounds 2 and 3. Making the board more permanent can save paper and preparation time in the future.

On the Day of the Activity

1. Arrange the desks or tables and chairs so there is a large open area for the drama that all students can see clearly. Place the three large signs on the wall to divide this area into three regions: outer space, the atmosphere, and the Earth.

2. Push tables or desks together so a group of four students can comfortably play the game together.

3. Set out the materials for the simulation game so they can easily be collected by a student from each group: 1 game board, 3 card sets, beans (12 of each color for each team), 3 score sheets, and 1 coin. Set the game boards to one side so you can hand them out separately during the demonstration that introduces the game.

4. Tape a game board to the chalkboard or wall to refer to when explaining the rules.

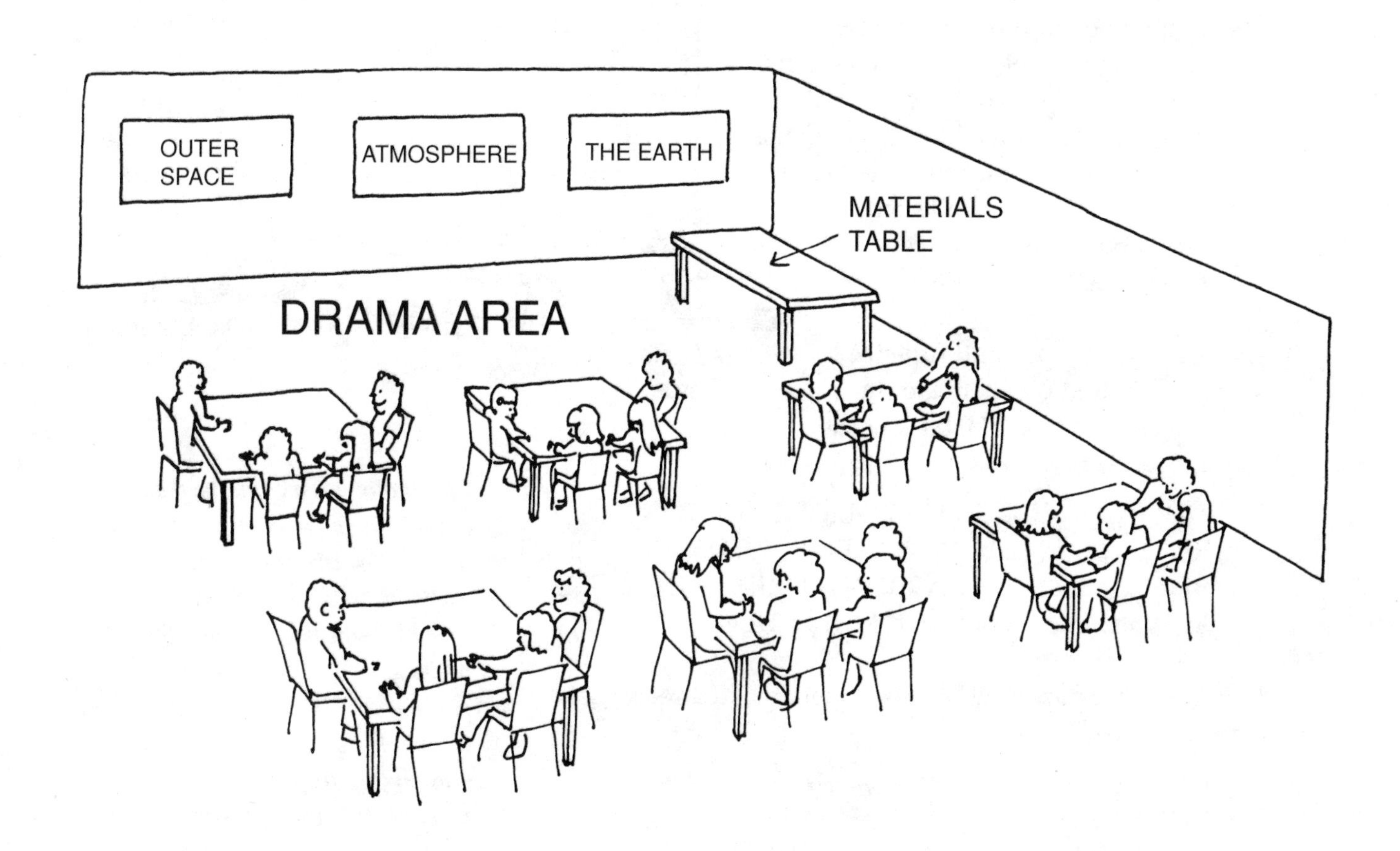

5. Draw a data table on the board, as follows:

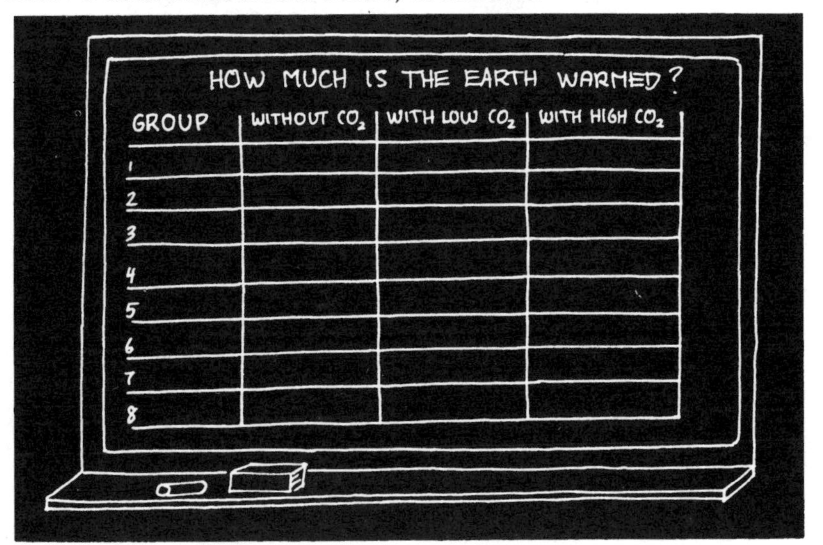

HOW MUCH IS THE EARTH WARMED?
GROUP
WITHOUT CO_2
WITH LOW CO_2
WITH HIGH CO_2
1
2
3
4
5
6
7
8

Reviewing The Greenhouse Experiment

1. Ask the students to summarize what they learned about the greenhouse effect during the previous session. Allow for some discussion, and help them summarize these main ideas:

   - Energy in the form of light and heat entered the bottle through the transparent plastic, where it heated up the ground and air.

   - Energy left the bottle through the transparent sides, and with warm air escaped through the open top.

   - When plastic was placed over the top, the warm air was trapped inside the bottle. It was prevented from mixing with the cool air outside, so the air heated up even more.

2. Explain to the students that, in the Earth's atmosphere, there is **no solid barrier** that prevents cool air and warm air from mixing. Instead, heat energy is trapped by **gases** in the Earth's atmosphere. This is another kind of "greenhouse effect." And, the gases that produce this warming are called ***greenhouse gases***. The way greenhouse gases trap heat is different from the way heat was trapped in the bottles during the experiments the students did in the previous session. Currently the greenhouse effect keeps the atmosphere at an average global temperature of 59°F.

Introducing the Global Warming Game

1. Tell the students they will play a game that illustrates how the greenhouse gases in the Earth's atmosphere trap heat.

2. Hand out a game board to each group of three to four students. Do not hand out the game materials yet. Explain that the demonstration they are about to see will show what happens in the game.

3. Turn on a light bulb from the previous session's experiments. Ask the students how they can tell when the light is turned on.
[They can see the light!]

4. Ask a volunteer seated near the front of the class to close his eyes and tell you when the light bulb comes close to his hand. Do this, and then ask how he could sense the light bulb.
[He could feel the warmth.]

5. Explain that scientists have discovered that the energy from a light bulb, or from the Sun, is composed of tiny packets of energy, called ***photons***. There are different sorts of photons, but two are especially important to know about for the game:
   1) photons we can *see* are called ***visible light photons***;
   2) photons we can *feel as warmth* are called ***infrared photons.***

You may wish to explain that visible light photons come in all colors of the rainbow (a chart showing the colors of the rainbow would be useful for this). A prism can be used to separate the stream of photons from a bulb, or the sun into all of the rainbow colors, called the ***spectrum****. A thermometer placed just beyond the visible red end of the spectrum shows that there are also invisible photons that carry heat energy. That is why these photons are called infrared.*

6. Tell the students that in the game visible light photons will be represented by white beans, and infrared photons will be represented by red beans—hold up one of each to demonstrate.

7. Briefly review the concept that all matter is made up of ***molecules***, and molecules are made of even smaller bits called ***atoms***. In gases and liquids, molecules are free to move around, while in solids, they are held together more rigidly.

8. Ask your students if they know what kinds of molecules make up the air.
   [76% nitrogen, 22% oxygen, 0.03% carbon dioxide, and other trace gases such as argon and helium]
   The dramatic demonstration and game the students are about to play helps show how carbon dioxide in the atmosphere affects global warming.

The Interaction Between Photons and Molecules

1. To begin the demonstration, point out the signs that divide the open area of the room into three regions: Outer Space, the Atmosphere, and the Earth. Tell the students to look at their game boards and see where these regions are located.

2. Ask for three volunteers.

   a. Ask one student to stand in **outer space**. Hand her a yellow sign that says "Visible Light Photon."

   b. Ask the second student to stand in the **atmosphere** area, and hand him a blue "Carbon Dioxide Molecule" sign.

   c. Ask the third student to stand in the Earth area. Hand her a gray or brown "Rock Molecule" sign.

   d. Ask all three volunteers to hold up their signs so the class can read them.

3. Ask the visible light photon carrier to slowly walk in a straight line past the carbon dioxide molecule in the atmosphere and approach the Earth.

4. As the visible light photon carrier does this, tell the class that the visible light photon originated in the Sun, and traveled in a straight line to the Earth, and is now entering and passing through the Earth's atmosphere.

5. Since carbon dioxide does not absorb visible light, the visible light photon passes right by the carbon dioxide molecule and approaches the Earth.

6. Ask the visible light photon carrier to stop at the Earth. Explain to the class that two things can happen when a visible light photon encounters Earth's surface. Either it will be reflected back into space, or absorbed by the Earth.

7. Point out the place on the game boards where this happens. Explain that, at this point in the game, each group will flip a coin to see whether the photon is absorbed or reflected. If the photon is reflected, it will go back into space.

8. Let's suppose the energy from the photon is absorbed. Tell the visible light photon carrier to hand his photon sign to the rock molecule and to stand by. The rock molecule will absorb this energy by vibrating. Ask the student to demonstrate this by jiggling her "Rock Molecule" sign.

9. Mention that, as your students know, you cannot see molecules moving in hot substances because the molecules are too small to see. However, if you touch the rock, your skin senses this vibration as warmth.

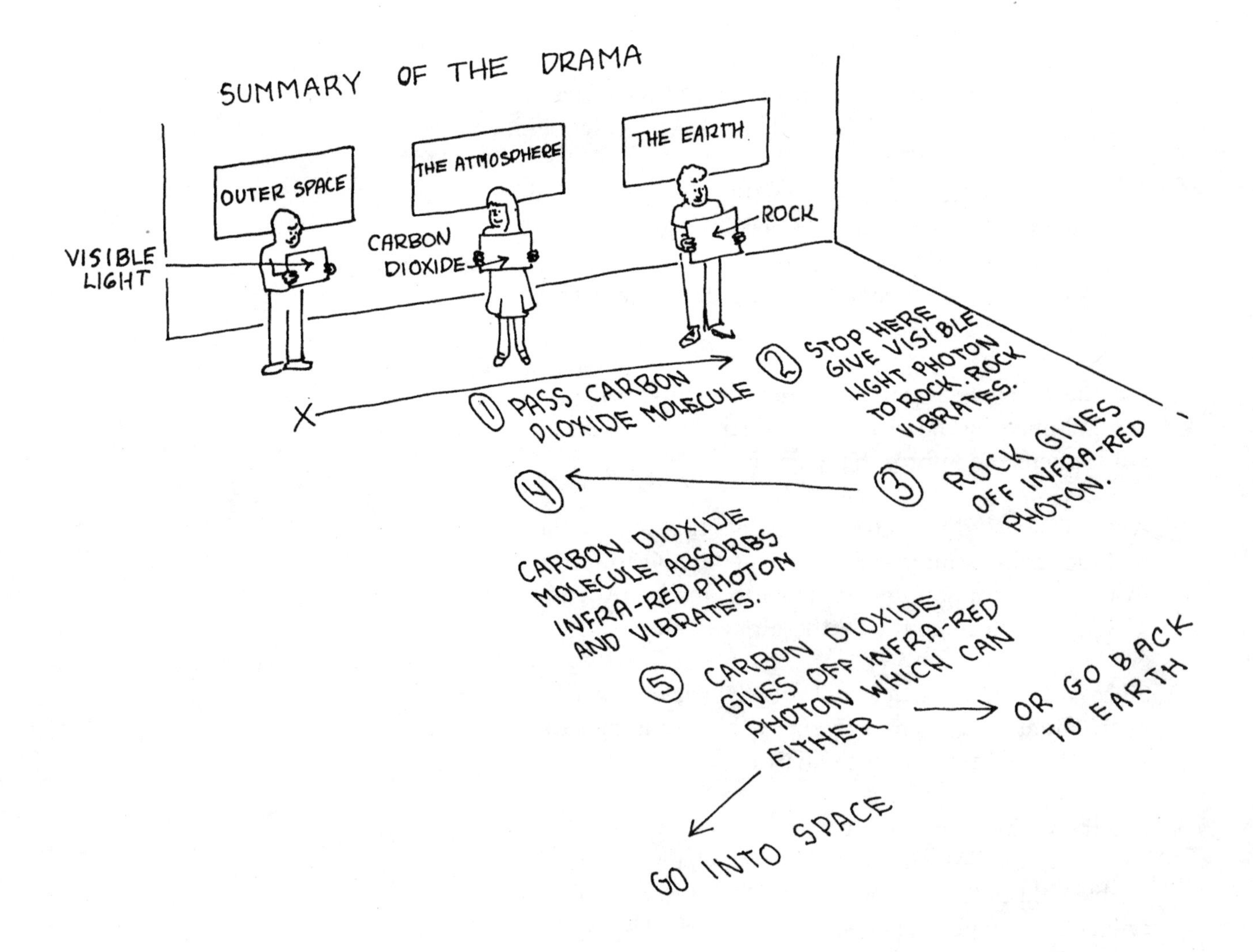

10. Explain that, after a while, the rock molecule will cool off, or stop vibrating. When it does, it gives up its vibrating energy by giving off an infrared photon. For example, if a rock has been in the sunlight all day, we can put our hands a few inches away from it and feel its warmth. What we feel are billions of infrared photons being given off by the rock molecules as they stop vibrating.

A brick, stone house, or concrete roadway that has been in the sun all day and still feels warm after the sun goes down might be a useful example that draws on the students' own experiences.

11. To demonstrate this, ask the rock molecule to stop vibrating and hand the sign that says "Infrared Photon" to the student who is standing nearby. The student who now holds the infrared photon then carries the energy away from the Earth.

12. Ask the infrared photon carrier to stop next to the carbon dioxide molecule on her trip away from the Earth.

13. Explain that carbon dioxide molecules absorb infrared photons easily. Ask, what should our two actors do when they come near each other?
[The photon carrier should hand her "photon" to the carbon dioxide molecule, who will vibrate.]
Ask, what should our actors do when the carbon dioxide molecule cools off?
[Stop vibrating, and hand the infrared photon sign back to the carrier.]

14. Explain to the class that, at this point, the photon might head off into space, or it might head back to the ground, where it might heat up another rock or another CO_2 molecule.

15. Ask the volunteers to take their seats and give the students a chance to ask questions about what happens when photons of energy from the Sun interact with matter.

Rules for the Global Warming Game

SUMMARY OF GAME

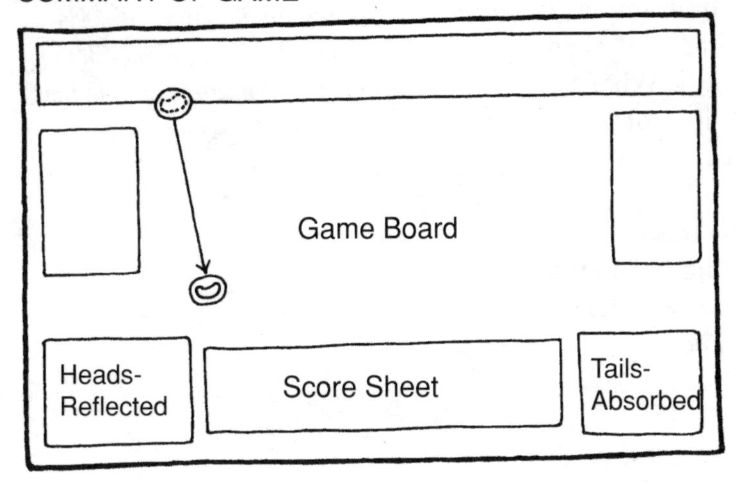

1. Push a photon through the atmosphere to the Earth.

2. Flip a coin. 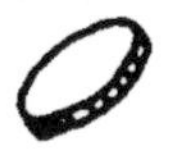

3. If **HEADS**, pick a "Heads-Reflected" card.

If **TAILS**, pick a "Tails-Absorbed" card.

4. Do what it says on the card.

5. When your photon leaves the atmosphere, remove it from the board.

The next person pushes the next photon, and so on.

1. Refer to the game board posted on the wall, and remind the students where outer space is, where the atmosphere is, and where the ground is.

2. To start the game, each group will place 12 white beans, representing visible light photons, in a row along the top. They will also receive two stacks of cards marked "Heads-Reflected" and "Tails-Absorbed." They should shuffle each deck separately and place it on the board in the correct space.

3. The players in each group will take turns to see what happens to one photon from the time it enters the Earth's atmosphere, to the time it finally exits into space. Taking the next photon in the row, the player will **push a photon** down to the Earth, **toss a coin** to see if it is reflected or absorbed by the Earth, and then **pick a card** from a pile determined by the coin toss (heads or tails) to see what happens next.

4. As in the demonstration the students just saw, if the visible light photon is absorbed it warms up the object that absorbs it. The object then cools off by emitting an infra-red photon. To represent this in the game, the player will replace the white bean with a red bean, and record a "W" on the score sheet provided for that round.

5. Each player continues to follow the photon he or she started with until the instructions on the card say that the photon goes off into space. At that point the photon is removed from the board, and the next student begins with the next photon in the row. Play for Round 1 continues until all 12 of the photons are removed from the board.

6. Tell the groups they are going to play the game three times.

   - In Round 1 the purpose is to see what happens on an imaginary planet where there is **only** oxygen and nitrogen in the atmosphere.

- In Round 2, they will **add some carbon dioxide** to the atmosphere, so it will be more like a real planet, such as Earth.

- In Round 3, they will play the game again, with **more carbon dioxide**, representing a future time when the Earth's atmosphere has more carbon dioxide.

7. The object of the game is to compare the amount of warming that occurs when more carbon dioxide is added to the atmosphere in Round 2 and Round 3.

Playing Round 1 of "The Global Warming Game"

1. Have a member from each group collect the materials for Round 1: 12 white beans, 12 red beans, one penny or other coin, two sets of cards ("Reflected" and "Absorbed"), and three score sheets (one for each round).

If a student asks what to place in the box marked "Carbon Dioxide Cards Go Here," explain that they will receive the cards later in the game.

2. Remind the students to arrange 12 white beans in the circles, across the top of the board, to shuffle the sets of cards separately, and place them on the board, and to place the "Round 1" score sheet on the board.

3. As groups finish setting up, remind them that the rules are: "Push a Photon — Toss a Coin — Pick a Card," and tell them to start playing the game.

4. As the groups start the game, make sure they are taking turns, and that each student follows a photon on its entire path from the time it enters the atmosphere to the time it leaves. Each student in a group of four will get three turns in the course of a round.

5. Also make sure that each group is reading and interpreting the cards correctly. Especially check that each group is recording a "W" on the score sheet each time a photon is absorbed by an object on the ground (as indicated on the cards).

Note: This is a good place to stop if you want to break the session into two class periods. It is preferable to hold the discussion shortly after the students have finished Round 3, rather than play the game entirely on one day, and then discuss the results a day or two later.

Playing Rounds 2 and 3

1. When all of the groups have finished Round 1, gain the attention of the entire class, and ask each group in turn how many times the absorption of photons had warmed the Earth in their game. Fill this number in for each group on the data table on the chalkboard.

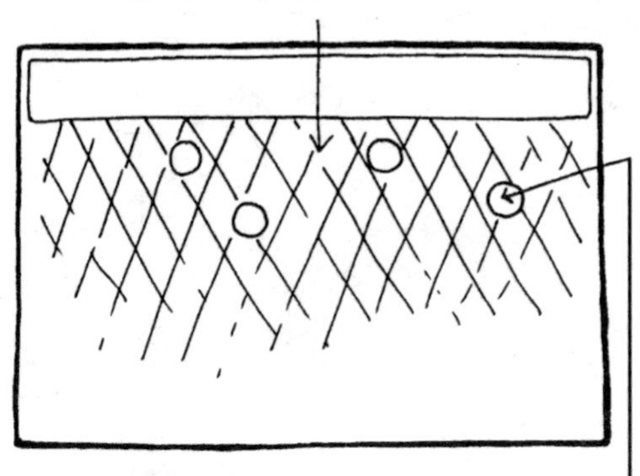

Use your coin to trace some circles scattered randomly in the atmosphere.

2. Announce that each group is now going to introduce carbon dioxide to the atmosphere in their game to see how this affects the warming of the Earth and atmosphere. Hand out the third set of cards, marked "Carbon Dioxide," and ask the students to shuffle the deck and place it on their game boards.

3. Use the posted copy of the game board to demonstrate how they should place three of their pennies randomly across the board, and trace around them in pencil, to make three circles to represent three carbon dioxide molecules in the atmosphere. They then should fill in the three circles with pencil.

4. Using a white bean, demonstrate how a visible light photon passes right through a carbon dioxide molecule on its way to Earth, or on its way into space after being reflected. Explain that visible light photons are only rarely absorbed by carbon dioxide, so in this game, **visible light will always pass through carbon dioxide molecules as though they weren't there.**

The rules of the game approximate the way actual photons behave. For example, about 45% of the visible light photons that make it to the Earth's surface are absorbed and another 25% are absorbed by the atmosphere. Also, visible light photons are not easily absorbed by carbon dioxide, whereas infrared photons are almost always absorbed.

5. Now ask the students to recall what happens when a visible light photon is absorbed by the Earth? [It warms the Earth, and is emitted as an infrared photon.]
   Tell the students that if an infrared photon encounters a carbon dioxide molecule, it is always absorbed. When this happens in the game, the player should write a "W" on the score sheet and pick up a card from the "Carbon Dioxide" pile.

6. Point out that the way to tell if an infrared photon runs into a carbon dioxide molecule in the game is to slide the exact middle of the bean sideways up the line. If any part of it touches a circle that represents a carbon dioxide molecule, the player should consider it absorbed, and pick up a "Carbon Dioxide" card.

7. Explain that carbon dioxide molecules cool off the way other molecules do—they give off an infrared photon. Sometimes those photons go into space, and sometimes they head back toward Earth. If they hit the Earth, they are always absorbed. This is shown in this round by simply picking a card from the "Tails-Absorbed" pile, without flipping a coin.

8. Ask if there are any questions about this part of the game. Remind students that, before beginning, they should place the score sheet for Round 2 on their game boards, and trace in three carbon dioxide molecules.

9. Tell the students that when they have finished Round 2, they should draw in three more carbon dioxide molecules and play the game again with the score sheet for Round 3. Let the games begin.

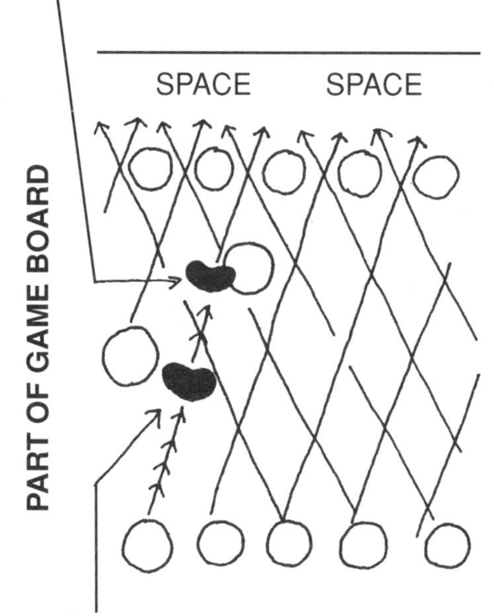
Photon is absorbed by carbon dioxide molecule.
SPACE
SPACE
PART OF GAME BOARD
Photon passes molecule of carbon dioxide.

Summarizing the Results

1. As groups finish their data collection for Rounds 2 and 3, have them record their data on the table on the chalkboard. They should write the total number of "W's" for each round in the appropriate boxes.

2. Have the groups who finish early play one more round with a mixture of carbon dioxide and methane in their "atmospheres." To play this round, tell them to use a quarter to trace a larger circle over three of the carbon dioxide circles. Tell them that these large circles represent a different gas, called "methane," which absorbs infrared photons even more easily than carbon dioxide. Their results can be added in an additional column on the chalkboard.

3. When all groups have completed at least Round 3, and the data table has been completed, collect all of the game boards and equipment, and focus the students' attention on the data.

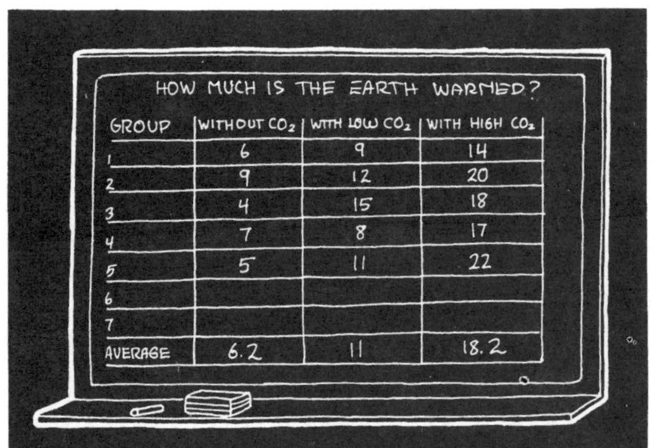

HOW MUCH IS THE EARTH WARMED?

GROUP	WITHOUT CO_2	WITH LOW CO_2	WITH HIGH CO_2
1	6	9	14
2	9	12	20
3	4	15	18
4	7	8	17
5	5	11	22
6			
7			
AVERAGE	6.2	11	18.2

4. Ask the class to suggest any trends they see in the results. [Warming without carbon dioxide will in general be lower than warming with low carbon dioxide, which in turn will be lower than warming in an atmosphere with a lot of carbon dioxide. Individual groups may have different results due to statistical variations.]

5. Ask the class to suggest how the results might be summarized to see trends, or patterns, in the data more clearly. [Add up or average the columns]
This can be done quickly with a calculator.

6. Tell the students that the rules of this game were not simply made up, but actually represent the way carbon dioxide molecules interact with visible light and infrared photons. Scientists use more precise games of this sort, usually played on a computer, to predict what will happen to the Earth's atmosphere when more carbon dioxide is added. It is called a ***simulation game.***

7. Ask the students, "How does carbon dioxide in the atmosphere warm the Earth?"
[Molecules of carbon dioxide absorb infrared photons, thus warming the atmosphere. When these molecules cool, they give off more infrared photons, some of which go back to the Earth's surface. Thus, some photons are absorbed several times before escaping into space.]

You can check students' understanding of the warming effect of carbon dioxide in the atmosphere by asking each of them to write their answers to these questions on a blank sheet of paper.

8. If students have difficulty making such generalizations, ask a series of focused questions, such as the following:

   a. When a photon is absorbed by a molecule, what happens to the molecule?
   [It jiggles around; it gets warmer.]

   b. What happens when that molecule cools off?
   [It gives off an infrared photon.]

   c. What happens to the infrared photon?
   [Sometimes it goes into space, sometimes it heads back to the Earth.]

   d. How many times can a photon warm the Earth when there is no carbon dioxide present?
   [Once]

e. How many times can a photon warm the Earth when there is carbon dioxide in the atmosphere?
[Several times]

9. Tell the students that carbon dioxide is called a "greenhouse gas" because it has the effect of trapping heat energy, or infrared photons, in the Earth's atmosphere. If it were not for the small amount of carbon dioxide in the atmosphere, the Earth would be in a continual Ice Age. Carbon dioxide helps maintain the Earth's temperature at a comfortably high level. Life on Earth has always depended on the greenhouse effect. The present concern is with the rapid increase in carbon dioxide due to human activity.

10. Ask the students to explain how this effect is different from the greenhouse effect they experimented with using plastic bottles.
[The plastic bottle greenhouses heated up because a solid barrier stopped the warm air from mixing with the cool air. In today's game, we found that carbon dioxide gas traps infrared photons so they can heat up many more molecules before they leave the Earth.]

You may want to emphasize again that it is the addition of greenhouse gases, not ozone depletion, that is the suggested cause of global warming. The only connection between these two problems is CFCs, which are causing the loss of ozone, and also absorb infrared photons. Reducing the use of these gases in industry could reduce both the loss of ozone and global warming, as long as the gases used to replace CFCs are not themselves greenhouse gases.

11. Explain that there are other gases in the atmosphere that absorb infrared photons even more readily than carbon dioxide. These gases, which include methane, nitrous oxide, and chlorofluorocarbons (CFCs), are also increasing rapidly due to human activity.

12. Tell the students that gases such as methane and CFCs can be represented by larger circles in the game. Ask the students what would happen if they played another round with larger circles.
[There would be more "W's" on the score sheet—more heat would be trapped.]
If some of the groups played an additional round with larger circles representing methane, ask them to report their results.

If your students ask why the temperature of the Earth has changed in the distant past, explain that there are several reasons. The most important is the Earth's changing orbit around the sun. There also have been periods of high volcanic activity which injected ash and carbon dioxide into the atmosphere.

13. Conclude the discussion by pointing out that carbon dioxide is considered to be the biggest problem because so much of it is produced every year.

Homework: The Past 160,000 Years

For more information about how we can tell what the temperature of the Earth was in the distant past, see page 126 in Behind the Scenes.

1. Hand out the two-page homework assignment, "Global Temperatures and Carbon Dioxide During the Past 160,000 Years" (see pages 56–57).

2. Discuss the graphs on page 57 with the students so they are able to interpret them correctly. For example, you might ask:

   - What is the vertical scale on the Temperature graph?
     [Degrees Fahrenheit]
     What is the horizontal scale?
     [Thousands of years before the present]

   - 160,000 years ago, how much colder was it than it is now?
     [15°F]

   - What is the vertical scale of the Carbon Dioxide graph?
     [Parts per million]
     What does "parts per million" mean?
     [If an average sample of 1 million molecules has only one molecule that is carbon dioxide, then the concentration of carbon dioxide is 1 part per million.]

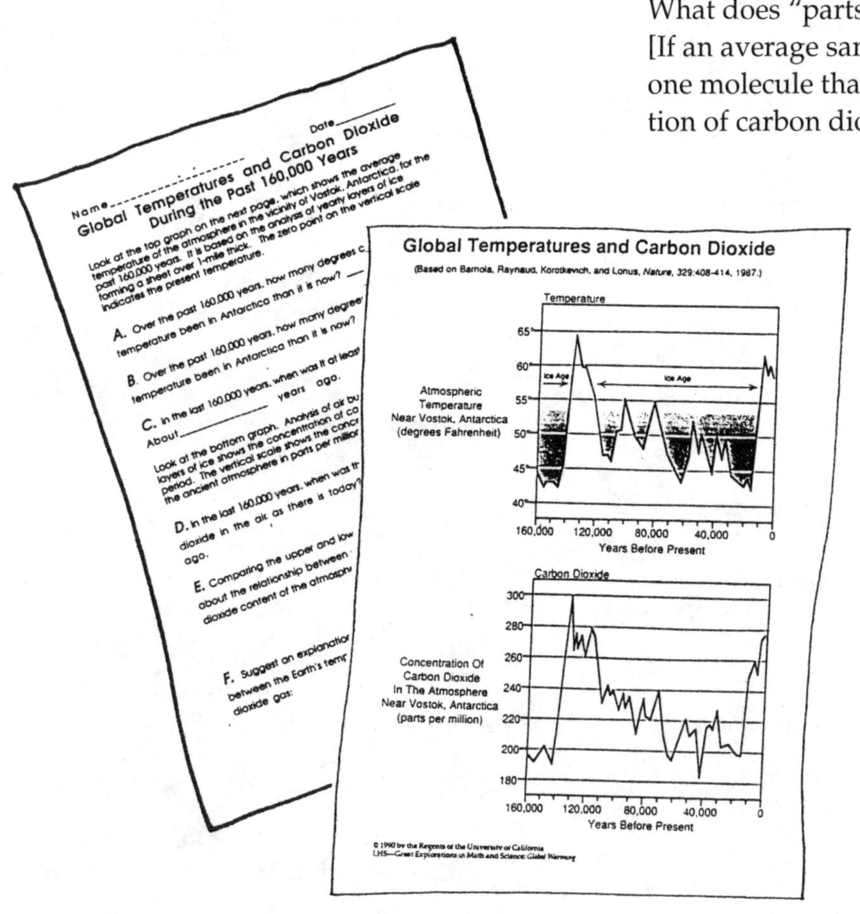

Global Temperatures and Carbon Dioxide During the Past 160,000 Years

Global Temperatures and Carbon Dioxide

(Based on Barnola, Raynaud, Korotkevich, and Lorius, *Nature*, 329:408-414, 1987.)

3. Tell students that in the next two sessions, they will perform laboratory experiments to determine the amount of carbon dioxide in different samples of air, such as the air in the room, and the air in their own breath.

Going Further

1. Encourage student groups, who have completed the three rounds of the game at different levels of carbon dioxide, to add "methane" to their atmospheres, by drawing larger circles on the boards in place of the carbon dioxide molecules. Then have students add other greenhouse gases to their boards, such as CFCs and nitrous oxides, in addition to carbon dioxide.

   Students could analyze articles or graphs that indicate the relative proportions of the various greenhouse gases in the atmosphere, and the effectiveness with which they absorb infrared radiation. The diameter of the circles, relative to the carbon dioxide already added, could be made to represent the relative absorptive capacity of these gases. The number of circles added could represent the concentrations of these gases in the atmosphere, relative to carbon dioxide. The game could then be played with a wide range of model atmospheres, to test out the likely effects of various greenhouse gases on the extent of global warming.

2. Replace the board simulation game with a class drama that illustrates the fate of photons as they enter the Earth's atmosphere and interact with matter in the same ways as in the game. Start with the drama activity that already precedes the game (showing what happens to a single photon that interacts with carbon dioxide or with a rock molecule on the Earth). Then set up a line of 12 students, representing 12 photons, and use the "Reflected," "Absorbed" and "Carbon Dioxide" cards from the game to determine what the students do. As in the game, play three rounds, with additional students representing carbon dioxide molecules in Rounds 2 and 3.

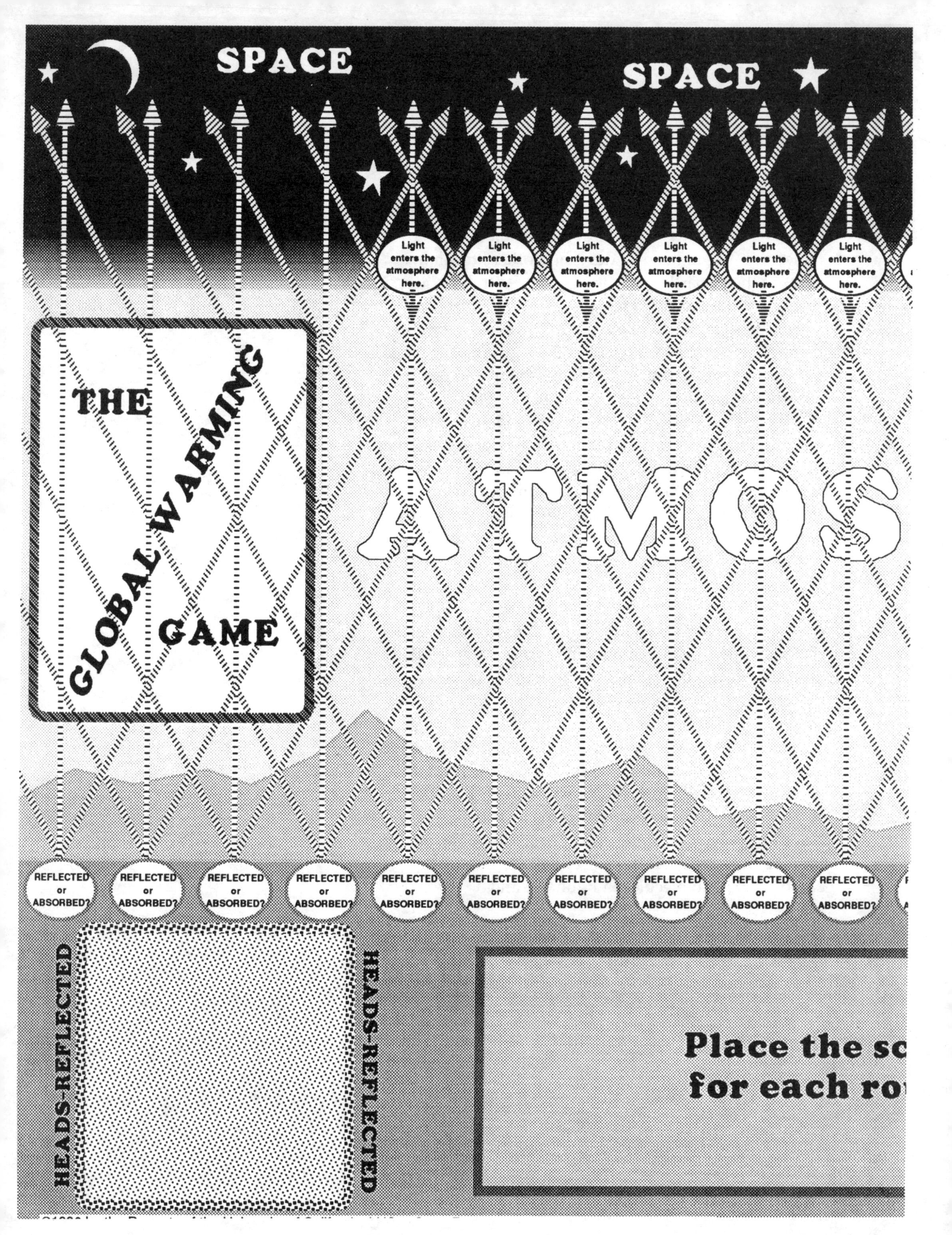
SPACE
SPACE
THE
GLOBAL WARMING
GAME
Light enters the atmosphere here.
Light enters the atmosphere here.
Light enters the atmosphere here.
Light enters the atmosphere here.
Light enters the atmosphere here.
Light enters the atmosphere here.
ATMOS
REFLECTED or ABSORBED?
REFLECTED or ABSORBED?
REFLECTED or ABSORBED?
REFLECTED or ABSORBED?
REFLECTED or ABSORBED?
REFLECTED or ABSORBED?
REFLECTED or ABSORBED?
REFLECTED or ABSORBED?
REFLECTED or ABSORBED?
REFLECTED or ABSORBED?
HEADS-REFLECTED
HEADS-REFLECTED
Place the sc
for each ro

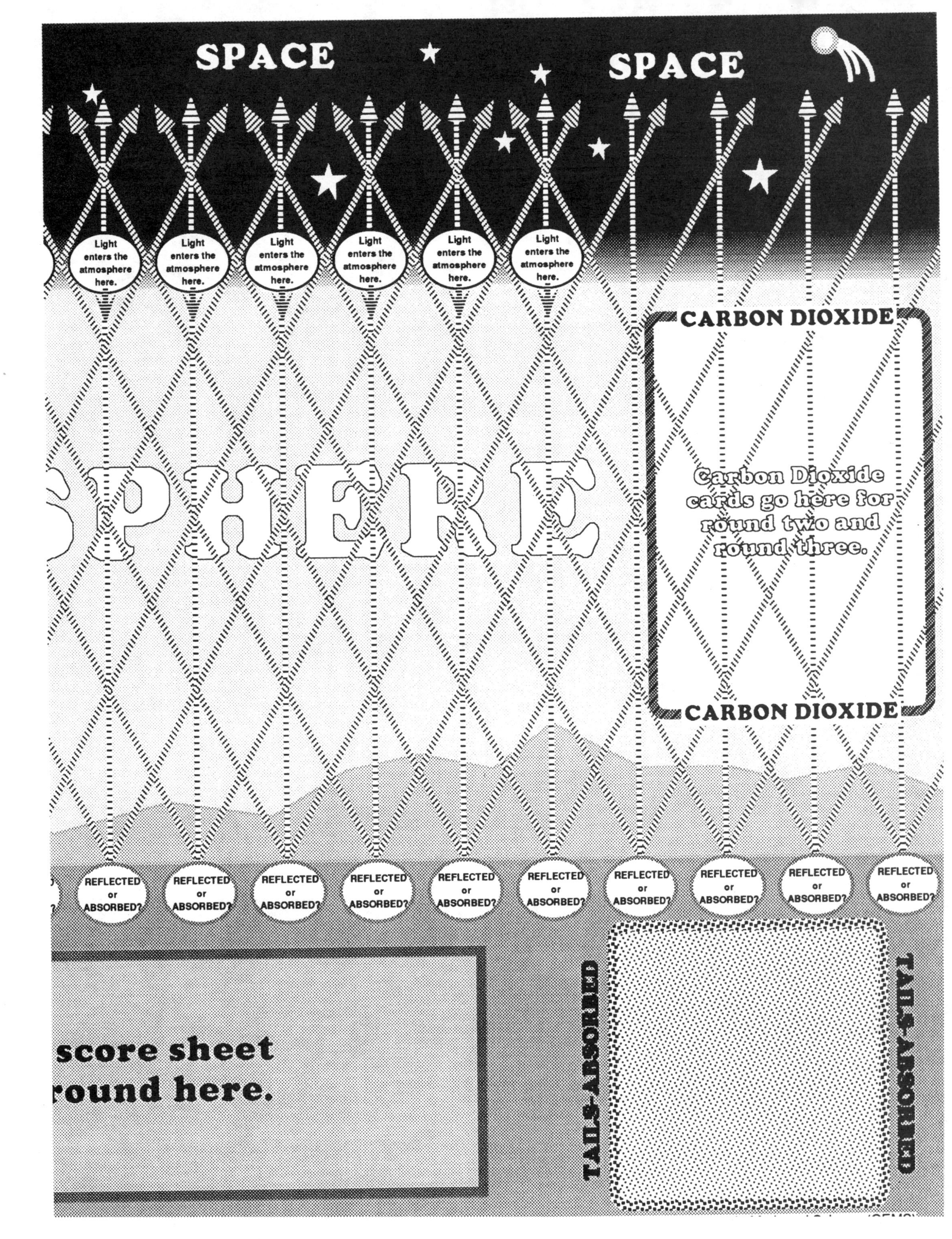
SPACE
SPACE
Light enters the atmosphere here.
Light enters the atmosphere here.
Light enters the atmosphere here.
Light enters the atmosphere here.
Light enters the atmosphere here.
Light enters the atmosphere here.
SPHERE
CARBON DIOXIDE
Carbon Dioxide cards go here for round two and round three.
CARBON DIOXIDE
REFLECTED or ABSORBED?
REFLECTED or ABSORBED?
REFLECTED or ABSORBED?
REFLECTED or ABSORBED?
REFLECTED or ABSORBED?
REFLECTED or ABSORBED?
REFLECTED or ABSORBED?
REFLECTED or ABSORBED?
REFLECTED or ABSORBED?
REFLECTED or ABSORBED?
score sheet
round here.
TAILS-ABSORBED
TAILS-ABSORBED

Score Sheets for
THE GLOBAL WARMING GAME

(CUT OUT EACH SHEET ALONG THE EDGE OF THE SHADED AREA

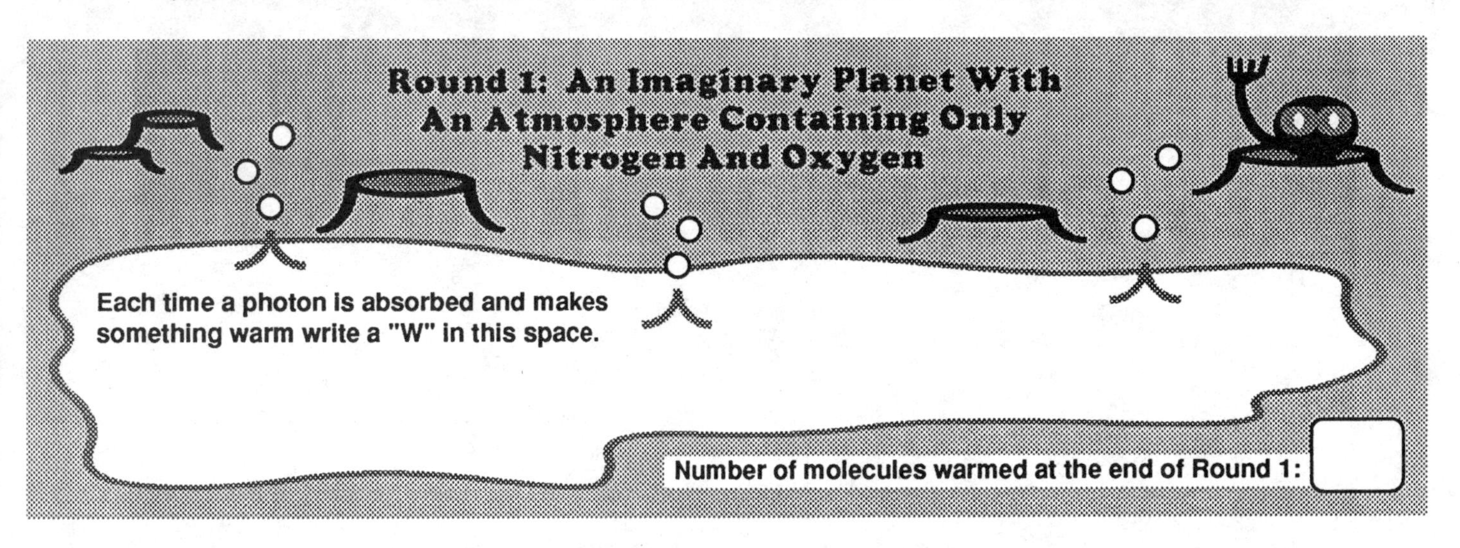
Round 1: An Imaginary Planet With
An Atmosphere Containing Only
Nitrogen And Oxygen
Each time a photon is absorbed and makes something warm write a "W" in this space.
Number of molecules warmed at the end of Round 1:

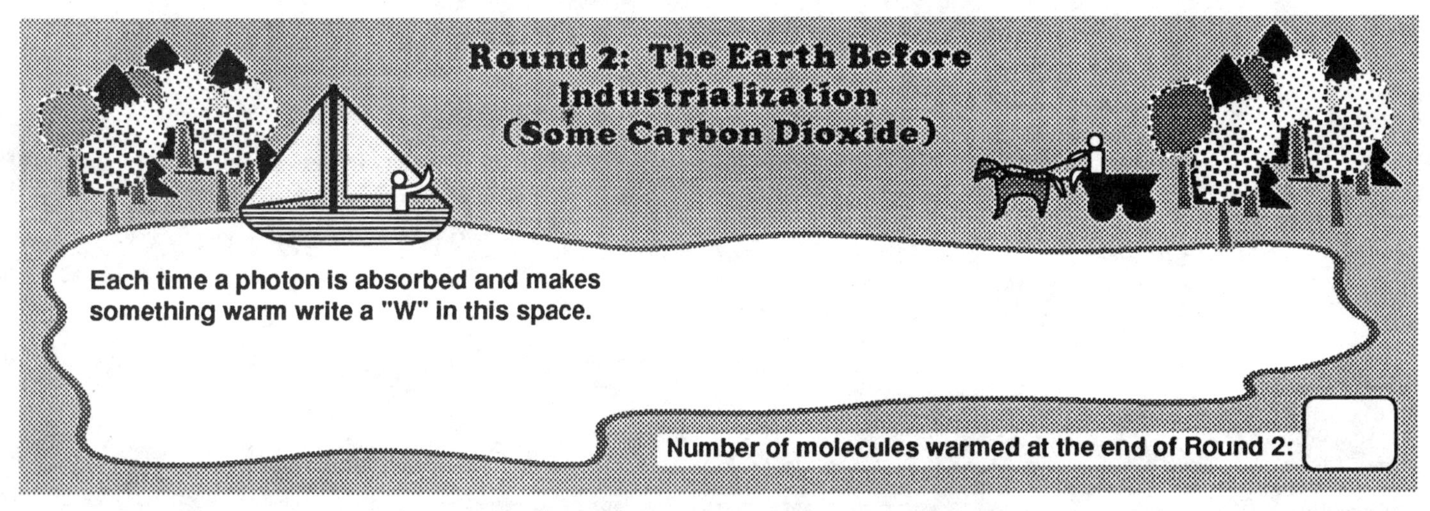
Round 2: The Earth Before
Industrialization
(Some Carbon Dioxide)
Each time a photon is absorbed and makes something warm write a "W" in this space.
Number of molecules warmed at the end of Round 2:

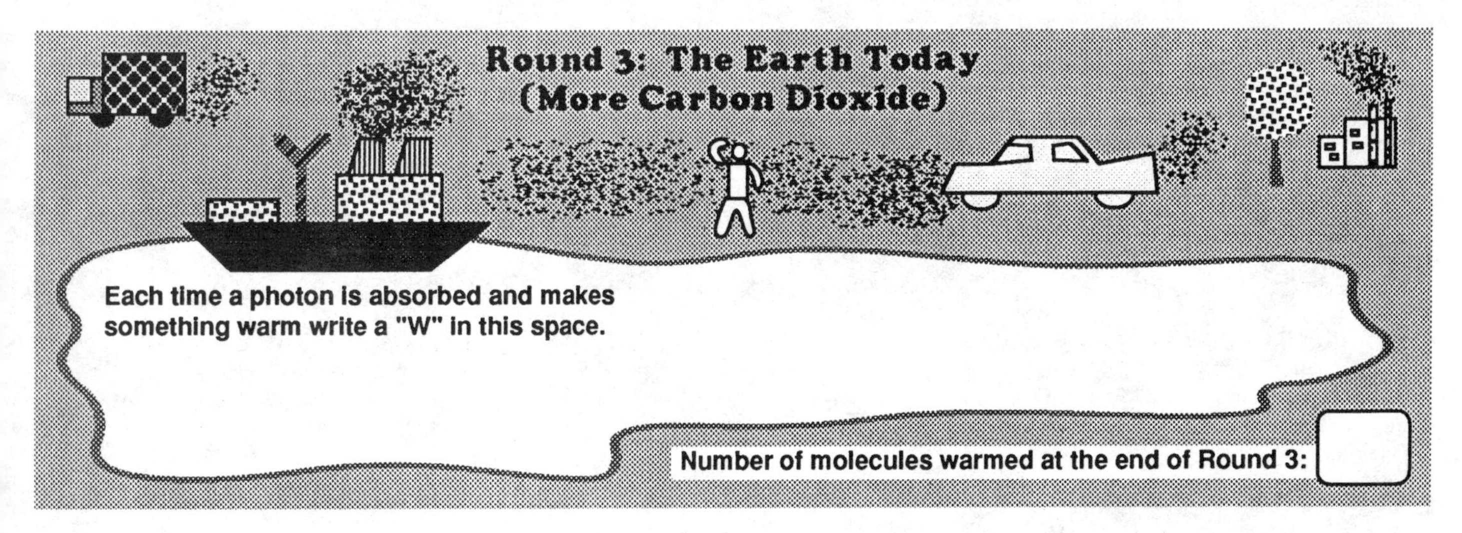
Round 3: The Earth Today
(More Carbon Dioxide)
Each time a photon is absorbed and makes something warm write a "W" in this space.
Number of molecules warmed at the end of Round 3:

ABSORBED

A paved road absorbs your photon and is warmed up a little bit.

✎ **Put a "W" on the scoresheet for this round.**

The road emits an *infra-red* photon. It goes upwards into space.

☞ **Use a red bean for your photon. Move it straight up away from the ground.**

ABSORBED

A leaf absorbs your photon and is warmed up a little bit.

✎ **Put a "W" on the scoresheet for this round.**

The leaf emits an *infra-red* photon. It goes sideways and hits another object.

☞ **Use a red bean for your photon. Move it to the right one space, and take another "ABSORBED" card.**

ABSORBED

A dark dress absorbs your photon and is warmed up a little bit.

✎ **Put a "W" on the scoresheet for this round.**

The dress emits an *infra-red* photon. It goes upwards into space.

☞ **Use a red bean for your photon. Move it away from the ground up and to the left.**

ABSORBED

A pine tree absorbs your photon and is warmed up a little bit.

✎ **Put a "W" on the scoresheet for this round.**

The tree emits an *infra-red* photon. It goes sideways and hits another object.

☞ **Use a red bean for your photon. Move it to the left one space, and take another "ABSORBED" card.**

ABSORBED

An oil slick absorbs your photon and is warmed up a little bit.

✎ **Put a "W" on the scoresheet for this round.**

The oil emits an *infra-red* photon. It goes upwards into space.

☞ **Use a red bean for your photon. Move it straight up away from the ground.**

ABSORBED

A roof tile absorbs your photon and is warmed up a little bit.

✎ **Put a "W" on the scoresheet for this round.**

The tile emits an *infra-red* photon. It goes upwards into space.

☞ **Use a red bean for your photon. Move it away from the ground up and to the right.**

ABSORBED

A stone absorbs your photon and is warmed up a little bit.

✎ **Put a "W" on the scoresheet for this round.**

The stone emits an *infra-red* photon. It goes upwards into space.

☞ **Use a red bean for your photon. Move it away from the ground up and to the right.**

ABSORBED

A gorilla absorbs your photon and is warmed up a little bit.

✎ **Put a "W" on the scoresheet for this round.**

The gorilla emits an *infra-red* photon. It goes upwards into space.

☞ **Use a red bean for your photon. Move it away from the ground up and to the left.**

ABSORBED

A snail absorbs your photon and is warmed up a little bit.

✎ **Put a "W" on the scoresheet for this round.**

The snail emits an *infra-red* photon. It goes sideways and hits another object.

☞ **Use a red bean for your photon. Move it to the right one space, and take another "ABSORBED" card.**

ABSORBED

A ripe plum absorbs your photon and is warmed up a little bit.

✎ **Put a "W" on the scoresheet for this round.**

The plum emits an *infra-red* photon. It goes upwards into space.

☞ **Use a red bean for your photon. Move it away from the ground up and to the left.**

ABSORBED

A lump of coal absorbs your photon and is warmed up a little bit.

✎ **Put a "W" on the scoresheet for this round.**

The coal emits an *infra-red* photon. It goes upwards into space.

☞ **Use a red bean for your photon. Move it away from the ground up and to the right.**

ABSORBED

Some mud absorbs your photon and is warmed up a little bit.

✎ **Put a "W" on the scoresheet for this round.**

The mud emits an *infra-red* photon. It goes upwards into space.

☞ **Use a red bean for your photon. Move it straight up away from the ground.**

ABSORBED

A <u>rose</u> absorbs your photon and is warmed up a little bit.

✎ **Put a "W" on the scoresheet for this round.**

The rose emits an *infra-red* photon. It goes upwards into space.

☞ **Use a red bean for your photon. Move it straight up away from the ground.**

ABSORBED

A <u>spider</u> absorbs your photon and is warmed up a little bit.

✎ **Put a "W" on the scoresheet for this round.**

The spider emits an *infra-red* photon. It goes upwards into space.

☞ **Use a red bean for your photon. Move it away from the ground up and to the right.**

ABSORBED

A <u>driver's seat</u> absorbs your photon and is warmed up a little bit.

✎ **Put a "W" on the scoresheet for this round.**

The seat emits an *infra-red* photon. It goes upwards into space.

☞ **Use a red bean for your photon. Move it away from the ground up and to the left.**

ABSORBED

a <u>mushroom</u> absorbs your photon and is warmed up a little bit.

✎ **Put a "W" on the scoresheet for this round.**

The mushrrom emits an *infra-red* photon. It goes upwards into space.

☞ **Use a red bean for your photon. Move it away from the ground up and to the left.**

ABSORBED

A <u>beach chair</u> absorbs your photon and is warmed up a little bit.

✎ **Put a "W" on the scoresheet for this round.**

The chair emits an *infra-red* photon. It goes sideways and hits another object.

☞ **Use a red bean for your photon. Move it to the left one space, and take another "ABSORBED" card.**

ABSORBED

A <u>grain of sand</u> absorbs your photon and is warmed up a little bit.

✎ **Put a "W" on the scoresheet for this round.**

The sand emits an *infra-red* photon. It goes upwards into space.

☞ **Use a red bean for your photon. Move it straight up away from the ground.**

ABSORBED

A <u>ice cream cone</u> absorbs your photon and is warmed up a little bit.

✎ **Put a "W" on the scoresheet for this round.**

The cone emits an *infra-red* photon. It goes upwards into space.

☞ **Use a red bean for your photon. Move it away from the ground up and to the right.**

ABSORBED

A <u>rock</u> absorbs your photon and is warmed up a little bit.

✎ **Put a "W" on the scoresheet for this round.**

The rock emits an *infra-red* photon. It goes sideways and hits another object.

☞ **Use a red bean for your photon. Move it to the right one space, and take another "ABSORBED" card.**

ABSORBED

A <u>lump of coal</u> absorbs your photon and is warmed up a little bit.

✎ **Put a "W" on the scoresheet for this round.**

The coal emits an *infra-red* photon. It goes upwards into space.

☞ **Use a red bean for your photon. Move it away from the ground up and to the left.**

ABSORBED

A <u>black cat</u> absorbs your photon and is warmed up a little bit.

✎ **Put a "W" on the scoresheet for this round.**

The cat emits an *infra-red* photon. It goes sideways and hits another object.

☞ **Use a red bean for your photon. Move it to the left one space, and take another "ABSORBED" card.**

ABSORBED

A <u>shoe</u> absorbs your photon and is warmed up a little bit.

✎ **Put a "W" on the scoresheet for this round.**

The shoe emits an *infra-red* photon. It goes upwards into space.

☞ **Use a red bean for your photon. Move it straight up away from the ground.**

ABSORBED

Some <u>grass</u> absorbs your photon and is warmed up a little bit.

✎ **Put a "W" on the scoresheet for this round.**

The grass emits an *infra-red* photon. It goes upwards into space.

☞ **Use a red bean for your photon. Move it away from the ground up and to the right.**

REFLECTED

Your photon has
hit a boulder.
It bounces to the right
where it hits another object.

☞ **Move one space to the right and toss a coin again to see whether you are absorbed or reflected. Pick a new card and follow its instructions.**

REFLECTED

Your photon has
hit a fishing boat.
It bounces up into space.

☞ **Move your photon straight up away from the ground.**

REFLECTED

Your photon has
hit a park bench.
It bounces up into space.

☞ **Move your photon away from the ground up and to the right.**

REFLECTED

Your photon has
hit a rock.
It bounces up into space.

☞ **Move your photon straight up away from the ground.**

REFLECTED

Your photon has
hit a cactus.
It bounces up into space.

☞ **Move your photon away from the ground up and to the right.**

REFLECTED

Your photon has
hit an ocean wave.
It bounces up into space.

☞ **Move your photon away from the ground up and to the left.**

REFLECTED

Your photon has
hit someone on the nose.
It bounces up into space.

☞ **Move your photon away from the ground up and to the right.**

REFLECTED

Your photon has
hit a mountain peak.
It bounces up into space.

☞ **Move your photon away from the ground up and to the left.**

REFLECTED

Your photon has
hit a sea gull.
It bounces to the right
where it hits another object.

☞ **Move one space to the right, and toss a coin again to see whether you are absorbed or reflected. Pick a new card and follow its instructions.**

REFLECTED

Your photon has
hit a grain of sand.
It bounces up into space.

☞ **Move your photon away from the ground up and to the left.**

REFLECTED

Your photon has
hit the side of a cliff.
It bounces to the right
where it hits another object.

☞ **Move one space to the right, and toss a coin again to see whether you are absorbed or reflected. Pick a new card and follow its instructions.**

REFLECTED

Your photon has
hit an iceberg.
It bounces up into space.

☞ **Move your photon straight up away from the ground.**

REFLECTED

Your photon has hit a tree trunk. It bounces to the right where it hits another object.

☞ **Move one space to the right, and toss a coin again to see whether you are absorbed or reflected. Pick a new card and follow its instructions.**

REFLECTED

Your photon has hit a lake. It bounces up into space.

☞ **Move your photon straight up away from the ground.**

REFLECTED

Your photon has hit a stone. It bounces up into space.

☞ **Move your photon away from the ground up and to the right.**

REFLECTED

Your photon has hit a sheet of ice. It bounces up into space.

☞ **Move your photon straight up away from the ground.**

REFLECTED

Your photon has hit a mouse. It bounces up into space.

☞ **Move your photon away from the ground up and to the right.**

REFLECTED

Your photon has hit a sidewalk. It bounces up into space.

☞ **Move your photon away from the ground up and to the left.**

REFLECTED

Your photon has hit a window sill. It bounces up into space.

☞ **Move your photon away from the ground up and to the right.**

REFLECTED

Your photon has hit a polar bear. It bounces up into space.

☞ **Move your photon away from the ground up and to the left.**

REFLECTED

Your photon has hit a pebble. It bounces to the right where it hits another object.

☞ **Move one space to the right, and toss a coin again to see whether you are absorbed or reflected. Pick a new card and follow its instructions.**

REFLECTED

Your photon has hit a patch of snow. It bounces up into space.

☞ **Move your photon away from the ground up and to the left.**

REFLECTED

Your photon has hit some sunglasses. It bounces to the right where it hits another object.

☞ **Move one space to the right and toss a coin again to see whether you are absorbed or reflected. Pick a new card and follow its instructions.**

REFLECTED

Your photon has hit a white hat. It bounces up into space.

☞ **Move your photon straight up away from the ground.**

Your photon has just hit a
CARBON DIOXIDE
molecule!

Check: If your photon is *visible light* (white bean) put this card back and continue in the same direction.

If your photon is *infra-red* (red bean) it is absorbed and the carbon dioxide is warmed up a little.

✎ **Put a "W" on the scoresheet for this round.**

The carbon dioxide emits an *infra-red* photon (use a red bean).

☞ **Follow the nearest path up and to the right.**

Your photon might go off into space (your turn is over), or it might hit another carbon dioxide (take a "CARBON DIOXIDE" card).

Your photon has just hit a
CARBON DIOXIDE
molecule!

Check: If your photon is *visible light* (white bean) put this card back and continue in the same direction.

If your photon is *infra-red* (red bean) it is absorbed and the carbon dioxide is warmed up a little.

✎ **Put a "W" on the scoresheet for this round.**

The carbon dioxide emits an *infra-red* photon (use a red bean).

☞ **Follow the nearest path straight down.**

Your photon might hit the ground (take an "ABSORBED" card), or it might hit another carbon dioxide (take a "CARBON DIOXIDE" card).

Your photon has just hit a
CARBON DIOXIDE
molecule!

Check: If your photon is *visible light* (white bean) put this card back and continue in the same direction.

If your photon is *infra-red* (red bean) it is absorbed and the carbon dioxide is warmed up a little.

✎ **Put a "W" on the scoresheet for this round.**

The carbon dioxide emits an *infra-red* photon (use a red bean).

☞ **Follow the nearest path up and to the left.**

Your photon might go off into space (your turn is over), or it might hit another carbon dioxide (take a "CARBON DIOXIDE" card).

Your photon has just hit a
CARBON DIOXIDE
molecule!

Check: If your photon is *visible light* (white bean) put this card back and continue in the same direction.

If your photon is *infra-red* (red bean) it is absorbed and the carbon dioxide is warmed up a little.

✎ **Put a "W" on the scoresheet for this round.**

The carbon dioxide emits an *infra-red* photon (use a red bean).

☞ **Follow the nearest path down and to the left.**

Your photon might hit the ground (take an "ABSORBED" card), or it might hit another carbon dioxide (take a "CARBON DIOXIDE" card).

Your photon has just hit a
CARBON DIOXIDE
molecule!

Check: If your photon is *visible light* (white bean) put this card back and continue in the same direction.

If your photon is *infra-red* (red bean) it is absorbed and the carbon dioxide is warmed up a little.

✎ **Put a "W" on the scoresheet for this round.**

The carbon dioxide emits an *infra-red* photon (use a red bean).

☞ **Follow the nearest path straight up .**

Your photon might go off into space (your turn is over), or it might hit another carbon dioxide (take a "CARBON DIOXIDE" card).

Your photon has just hit a
CARBON DIOXIDE
molecule!

Check: If your photon is *visible light* (white bean) put this card back and continue in the same direction.

If your photon is *infra-red* (red bean) it is absorbed and the carbon dioxide is warmed up a little.

✎ **Put a "W" on the scoresheet for this round.**

The carbon dioxide emits an *infra-red* photon (use a red bean).

☞ **Follow the nearest path down and to the right.**

Your photon might hit the ground (take an "ABSORBED" card), or it might hit another carbon dioxide (take a "CARBON DIOXIDE" card).

Your photon has just hit a
CARBON DIOXIDE
molecule!

Check: If your photon is *visible light* (white bean) put this card back and continue in the same direction.

If your photon is *infra-red* (red bean) it is absorbed and the carbon dioxide is warmed up a little.

✎ **Put a "W" on the scoresheet for this round.**

The carbon dioxide emits an *infra-red* photon (use a red bean).

☞ **Follow the nearest path up and to the left.**

Your photon might go off into space (your turn is over), or it might hit another carbon dioxide (take a "CARBON DIOXIDE" card).

Your photon has just hit a
CARBON DIOXIDE
molecule!

Check: If your photon is *visible light* (white bean) put this card back and continue in the same direction.

If your photon is *infra-red* (red bean) it is absorbed and the carbon dioxide is warmed up a little.

✎ **Put a "W" on the scoresheet for this round.**

The carbon dioxide emits an *infra-red* photon (use a red bean).

☞ **Follow the nearest path straight down.**

Your photon might hit the ground (take an "ABSORBED" card), or it might hit another carbon dioxide (take a "CARBON DIOXIDE" card).

Your photon has just hit a
CARBON DIOXIDE
molecule!

Check: If your photon is *visible light* (white bean) put this card back and continue in the same direction.

If your photon is *infra-red* (red bean) it is absorbed and the carbon dioxide is warmed up a little.

✎ **Put a "W" on the scoresheet for this round.**

The carbon dioxide emits an *infra-red* photon (use a red bean).

☞ **Follow the nearest path up and to the right.**

Your photon might go off into space (your turn is over), or it might hit another carbon dioxide (take a "CARBON DIOXIDE" card).

Your photon has just hit a
CARBON DIOXIDE
molecule!

Check: If your photon is *visible light* (white bean) put this card back and continue in the same direction.

If your photon is *infra-red* (red bean) it is absorbed and the carbon dioxide is warmed up a little.

✎ **Put a "W" on the scoresheet for this round.**

The carbon dioxide emits an *infra-red* photon (use a red bean).

☞ **Follow the nearest path down and to the right.**

Your photon might hit the ground (take an "ABSORBED" card), or it might hit another carbon dioxide (take a "CARBON DIOXIDE" card).

Your photon has just hit a
CARBON DIOXIDE
molecule!

Check: If your photon is *visible light* (white bean) put this card back and continue in the same direction.

If your photon is *infra-red* (red bean) it is absorbed and the carbon dioxide is warmed up a little.

✎ **Put a "W" on the scoresheet for this round.**

The carbon dioxide emits an *infra-red* photon (use a red bean).

☞ **Follow the nearest path straight up.**

Your photon might go off into space (your turn is over), or it might hit another carbon dioxide (take a "CARBON DIOXIDE" card).

Your photon has just hit a
CARBON DIOXIDE
molecule!

Check: If your photon is *visible light* (white bean) put this card back and continue in the same direction.

If your photon is *infra-red* (red bean) it is absorbed and the carbon dioxide is warmed up a little.

✎ **Put a "W" on the scoresheet for this round.**

The carbon dioxide emits an *infra-red* photon (use a red bean).

☞ **Follow the nearest path down and to the left.**

Your photon might hit the ground (take an "ABSORBED" card), or it might hit another carbon dioxide (take a "CARBON DIOXIDE" card).

Your photon has just hit a
CARBON DIOXIDE
molecule!

Check: If your photon is *visible light* (white bean) put this card back and continue in the same direction.

If your photon is *infra-red* (red bean) it is absorbed and the carbon dioxide is warmed up a little.

✎ **Put a "W" on the scoresheet for this round.**

The carbon dioxide emits an *infra-red* photon (use a red bean).

☞ **Follow the nearest path up and to the left.**

Your photon might go off into space (your turn is over), or it might hit another carbon dioxide (take a "CARBON DIOXIDE" card).

Your photon has just hit a
CARBON DIOXIDE
molecule!

Check: If your photon is *visible light* (white bean) put this card back and continue in the same direction.

If your photon is *infra-red* (red bean) it is absorbed and the carbon dioxide is warmed up a little.

✎ **Put a "W" on the scoresheet for this round.**

The carbon dioxide emits an *infra-red* photon (use a red bean).

☞ **Follow the nearest path straight down.**

Your photon might hit the ground (take an "ABSORBED" card), or it might hit another carbon dioxide (take a "CARBON DIOXIDE" card).

Your photon has just hit a
CARBON DIOXIDE
molecule!

Check: If your photon is *visible light* (white bean) put this card back and continue in the same direction.

If your photon is *infra-red* (red bean) it is absorbed and the carbon dioxide is warmed up a little.

✎ **Put a "W" on the scoresheet for this round.**

The carbon dioxide emits an *infra-red* photon (use a red bean).

☞ **Follow the nearest path up and to the right.**

Your photon might go off into space (Your turn is over), or it might hit another carbon dioxide (take a "CARBON DIOXIDE" card).

Your photon has just hit a
CARBON DIOXIDE
molecule!

Check: If your photon is *visible light* (white bean) put this card back and continue in the same direction.

If your photon is *infra-red* (red bean) it is absorbed and the carbon dioxide is warmed up a little.

✎ **Put a "W" on the scoresheet for this round.**

The carbon dioxide emits an *infra-red* photon (use a red bean).

☞ **Follow the nearest path down and to the left.**

Your photon might hit the ground (take an "ABSORBED" card), or it might hit another carbon dioxide (take a "CARBON DIOXIDE" card).

Your photon has just hit a
CARBON DIOXIDE
molecule!

Check: If your photon is *visible light* (white bean) put this card back and continue in the same direction.

If your photon is *infra-red* (red bean) it is absorbed and the carbon dioxide is warmed up a little.

✎ **Put a "W" on the scoresheet for this round.**

The carbon dioxide emits an *infra-red* photon (use a red bean).

☞ **Follow the nearest path straight up.**

Your photon might go off into space (your turn is over), or it might hit another carbon dioxide (take a "CARBON DIOXIDE" card).

Your photon has just hit a
CARBON DIOXIDE
molecule!

Check: If your photon is *visible light* (white bean) put this card back and continue in the same direction.

If your photon is *infra-red* (red bean) it is absorbed and the carbon dioxide is warmed up a little.

✎ **Put a "W" on the scoresheet for this round.**

The carbon dioxide emits an *infra-red* photon (use a red bean).

☞ **Follow the nearest path down and to the right.**

Your photon might hit the ground (take an "ABSORBED" card), or it might hit another carbon dioxide (take a "CARBON DIOXIDE" card).

Sources of Greenhouse Gases from Human Activity

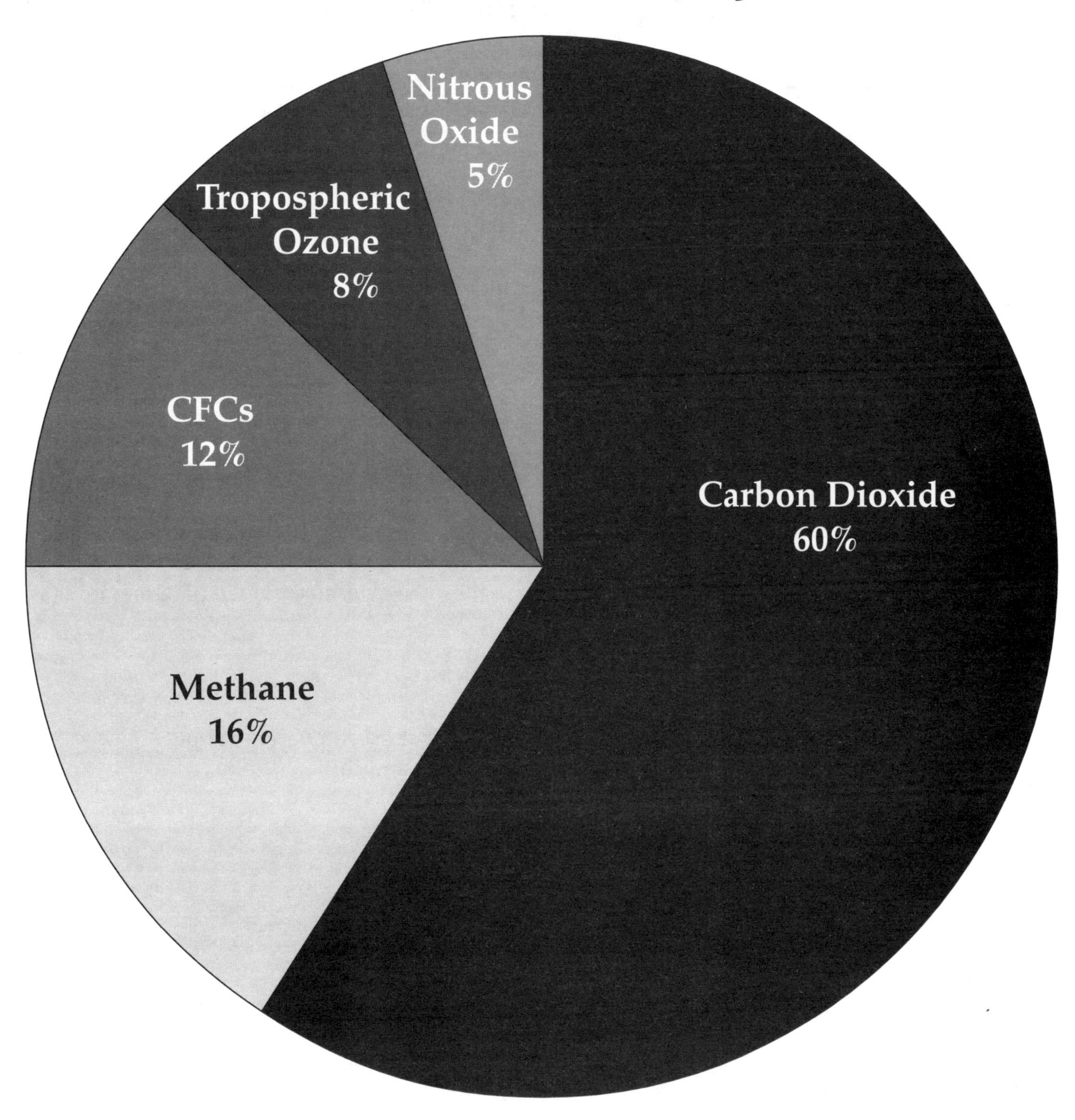

Source: Graedel, T.E., and Crutzen, P.J., 1990, The Changing Atmosphere, *in* Managing Planet Earth, readings from Scientific American magazine: New York, W.H. Freeman and Co., p.13–23.

Name ______________________________ Date ______________

Global Temperatures and Carbon Dioxide During the Past 160,000 Years

Look at the top graph on the next page, which shows the average global temperature for the past 160,000 years. It is based on the analysis of yearly layers of ice forming a sheet over one mile thick.

A. Over the past 160,000 years, how many degrees colder has the temperature been than it is now? ____________°F.

B. Over the past 160,000 years, how many degrees warmer has the temperature been than it is now? ____________°F.

C. In the last 160,000 years, when was it at least as warm as it is today? About ____________ years ago.

D. How much has the Earth warmed since 1850? About ____________ °F.

Look at the bottom graph. Analysis of air bubbles trapped in the same layers of ice shows the concentration of carbon dioxide during the same period. The vertical scale shows the concentration of carbon dioxide in the ancient atmosphere in parts per million.

E. When was the last time there was as much carbon dioxide in the air as there was in 1850? About ____________ years ago.

F. Comparing the upper and lower graphs, what do you observe about the relationship between global temperature and carbon dioxide content of the atmosphere during the past 160,000 years?

G. What year has the greatest concentration of carbon dioxide gas? ____________

H. Suggest an explanation for the relationship you observe between the Earth's temperature and the concentration of carbon dioxide gas.

Global Temperatures and Carbon Dioxide

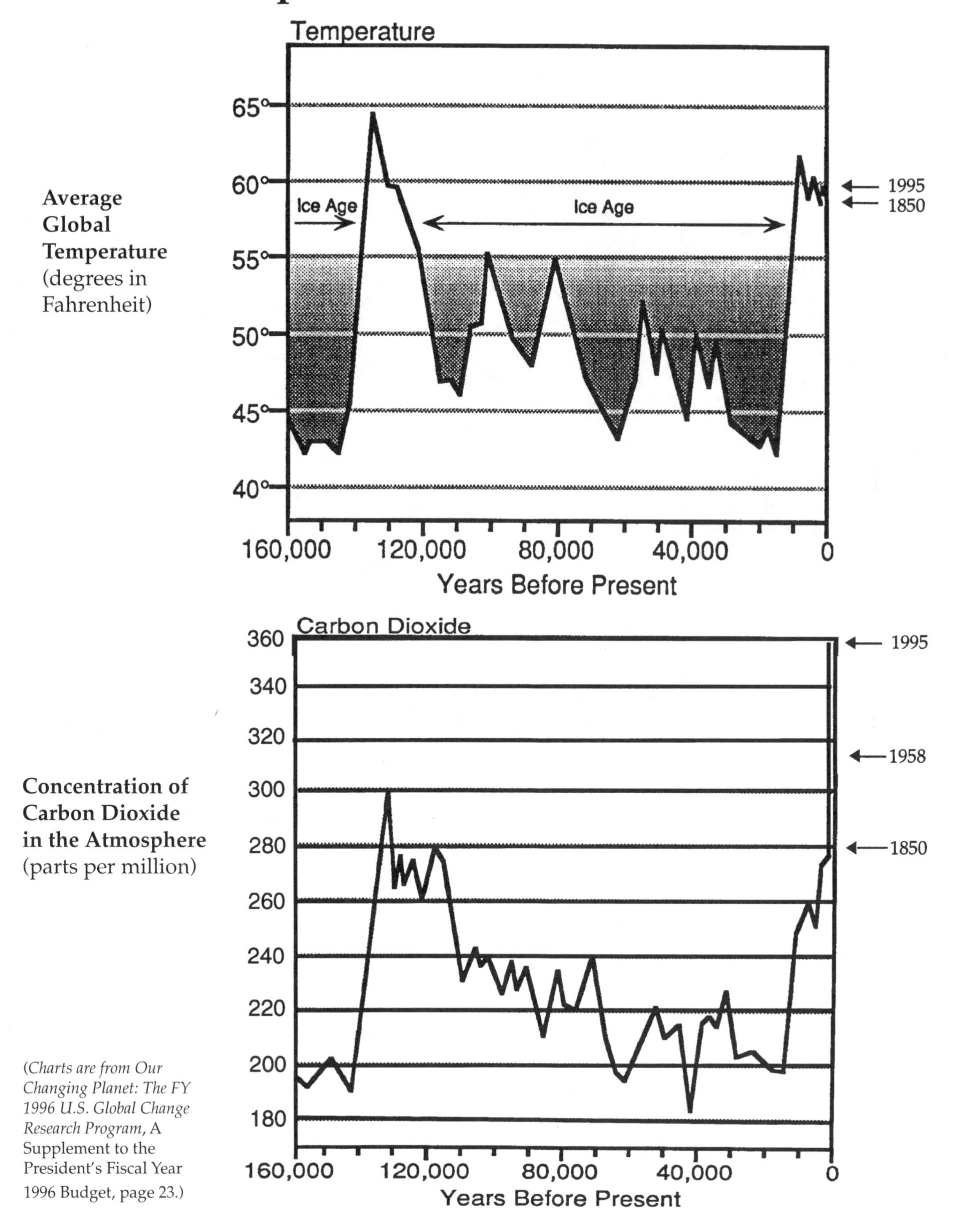

(*Charts are from Our Changing Planet: The FY 1996 U.S. Global Change Research Program*, A Supplement to the President's Fiscal Year 1996 Budget, page 23.)

Session 4: Detecting Carbon Dioxide

Overview

In the previous session, the students learned that many scientists think increased carbon dioxide levels in the atmosphere will cause the Earth's climate to become warmer. That naturally raises the question, "Where is the carbon dioxide coming from?" In Session 4, the students learn a technique for detecting the presence of carbon dioxide in a sample of gas. The technique they learn in this session will be applied in Session 5 to compare the concentrations of carbon dioxide in human breath and auto exhaust.

After reviewing the results of the previous sessions and discussing the homework, the students observe a demonstration of how to generate carbon dioxide, and learn its properties. The students learn how to determine the level of carbon dioxide in gas samples, using bromothymol blue (BTB) solution. The class ends with a discussion of the sources of carbon dioxide in the atmosphere.

The purposes of this session are to:

(1) familiarize the students with the properties of carbon dioxide;

(2) provide practice in using a technique for detecting carbon dioxide;

(3) prepare students for conducting a more thorough investigation of some common sources of carbon dioxide in Session 5.

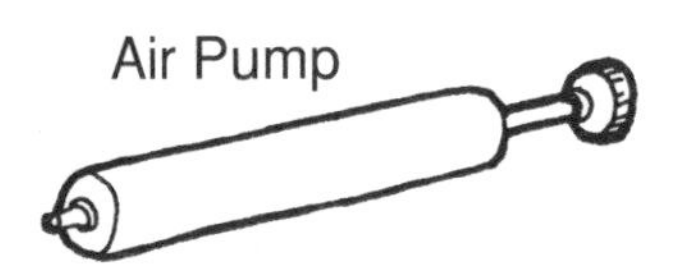

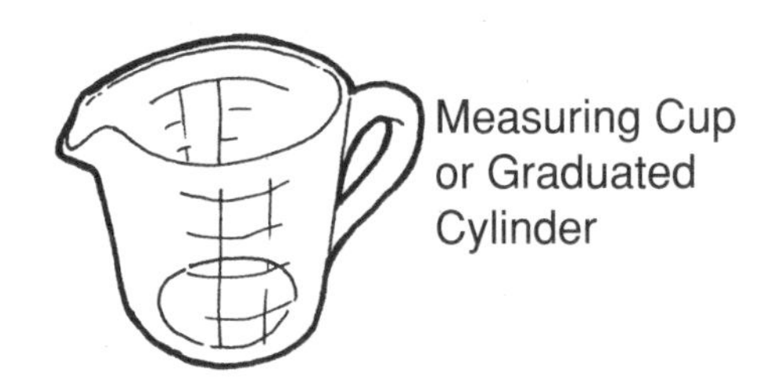

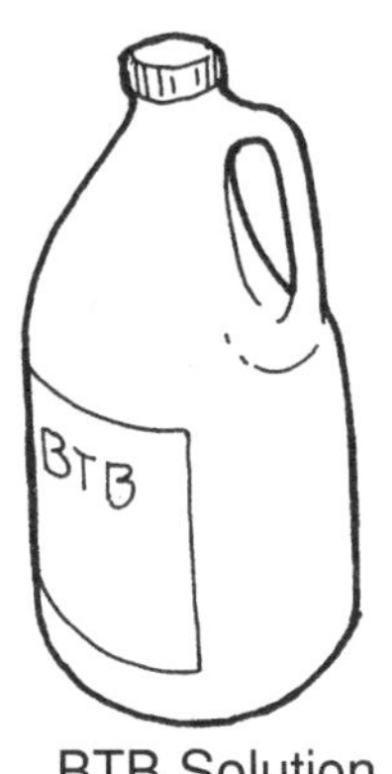

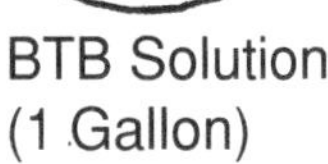

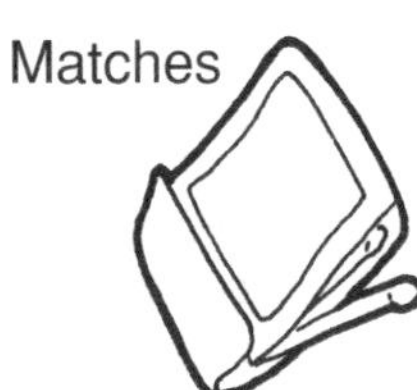

What You Need

For the class

- ❑ 1 box of baking soda
- ❑ 1 gallon of diluted Bromothymol Blue Solution (BTB), see page 61 for how to prepare
- ❑ 1 air pump, either bicycle pump or balloon inflator
- ❑ 1 32 oz. bottle of white vinegar
- ❑ 1 graduated cylinder or measuring cup
- ❑ 1 short candle (about 2" tall)
- ❑ 1 open clear glass or plastic container with sides taller than the candle
- ❑ 1 chalkboard or butcher paper and marker
- ❑ 1 gallon of white vinegar (32 oz. for each class session)

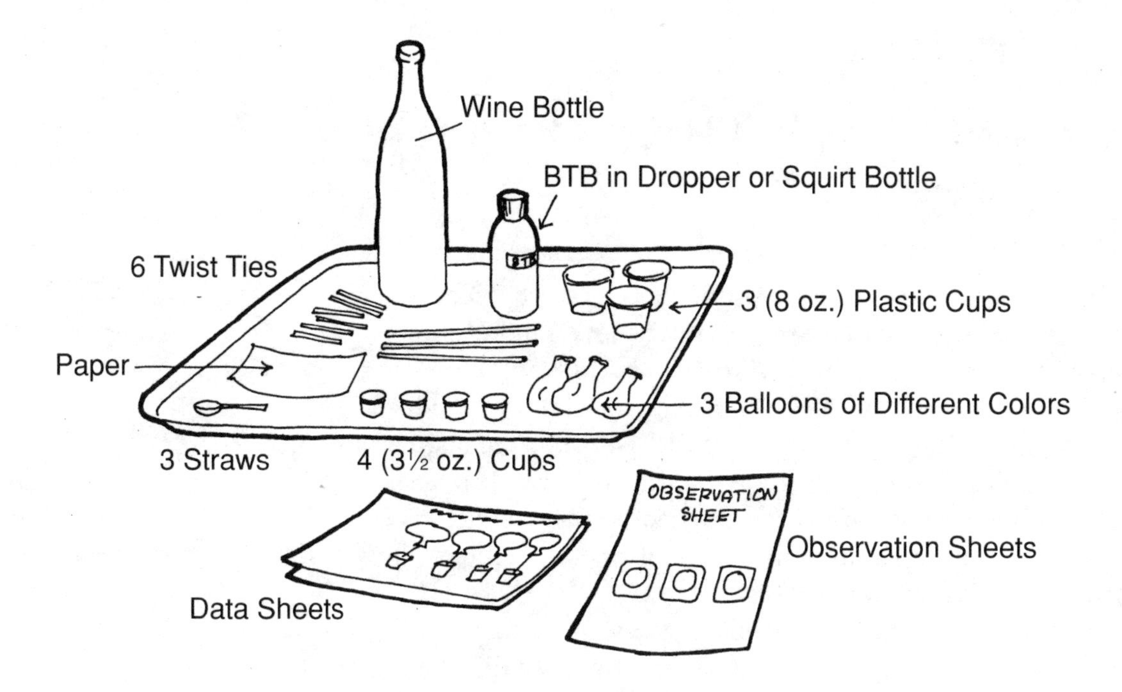

For each group of 4 students

An additional number of the items marked with an asterisk will be required in Session 5 (see pages 78–79).

- ❑ 1 empty glass wine bottle, 750 ml
- ❑ 1 dropper or squirt bottle, about 6 oz. size
- ❑ 1 sheet of white paper
- ❑ 1 teaspoon
- *❑ 3 plastic cups, 8 oz. size
- *❑ 4 small clear plastic cups (such as graduated medicine cups or 3½ oz. clear Solo® cups)
- *❑ 3 balloons, 8–10" in diameter, in three different colors
- *❑ 3 plastic straws
- *❑ 6 twist ties
- ❑ 1 handout: "Air and Carbon Dioxide Observation Sheet" (see page 74)
- ❑ 1 roll of masking tape or other object with a hole about 3½" diameter; same size for all groups
- ❑ 1 tray

optional

- ❑ 1 graduated cylinder

For each student

- ❑ 1 handout: "Air and Carbon Dioxide Data Sheet" (see page 75)
- ❑ 1 homework sheet: "Surprise Increase in Atmosphere's CO_2" (see pages 72–73)

IMPORTANT NOTE
When cleaning bottles, simply rinse with water. Do NOT use soap! (Soap residue will bubble and will affect the pH of the gas test.) If soap has been used inadvertently, make sure the bottles are extremely well rinsed out with water.

Getting Ready

Several Weeks Before Beginning this Unit

Bromothymol blue solution (BTB), used here to test for the presence and relative concentration of carbon dioxide, is an acid/base indicator that measures the pH range from 6.0 (yellow) through green to 7.6 (blue). The blue BTB solution moves toward the yellow (or more acidic) end of the indicator range with increasing concentrations of CO_2 due to the carbonic acid created when the CO_2 is bubbled through water.

1. Collect the glass wine bottles. Other glass or plastic bottles may be substituted, as long as they are about as large and tall as a standard-size wine bottle, and the neck of a balloon can fit over the top with an airtight seal. Two-liter soda bottles usually do not permit a good seal at the top, but other bottles may work well. Test bottles by using one to fill a balloon with carbon dioxide as described in this activity.

2. Obtain bromothymol blue (BTB) solution from a high school science lab or chemical supply house. BTB is available in either concentrated liquid (aqueous) or powdered form. The liquid form is easier to use.

Before the Day of the Activity

BTB is a much-used chemical indicator and is not considered dangerous given normal precautions and conditions. However, BTB should be handled using standard lab safety procedures. When preparing the solution, protective gloves and goggles are recommended. While there is little data available on adverse health effects, spills of BTB on bare skin may be irritating, and the affected area should be washed immediately with soap or mild detergent and large amounts of water. If the eyes are affected, they also should be washed out with large amounts of water. If BTB is swallowed, it may cause gastrointestinal irritation. Do not induce vomiting; get medical attention immediately.

1. Make a one gallon solution of BTB and water, which will be enough for one class for Sessions 4 and 5. If beginning with the concentrated liquid, fill a gallon bottle 9⁄10ths full with tap water, and add BTB concentrate until the solution is a deep, transparent blue. The exact concentration is not critical. However, you should test the solution by pouring 15 ml (½ oz.) BTB solution into one of the small clear cups. Using a straw, bubble one lungful of breath through the small cup of solution. If the solution turns green, it is OK. If it stays blue, or only slightly bluish green, it is too concentrated. Pour out some solution, add more water and test again. If it turns yellow, it is too diluted, and you need to add more BTB.

   If you start with BTB in powdered form, the instructions are the same, except you must first prepare the concentrate. Put 1 gram of BTB powder into a 1 liter container. Add 16 ml of 0.1 molar sodium hydroxide (0.1 M NaOH) and dissolve the crystals of BTB. Add 1 liter of water to make 1 liter of concentrate. For each group of four students, fill a dropper or squirt bottle with about 6 oz. of prepared BTB solution.

2. Make copies of the observation sheet "Air and Carbon Dioxide Observation Sheet" (1 per group, see page 74), the data recording sheet "Air and Carbon Dioxide Data Sheet" (1 per student, see page 75), and the homework sheet "Surprise Increase in Atmosphere's CO_2" (1 per student, see pages 72–73).

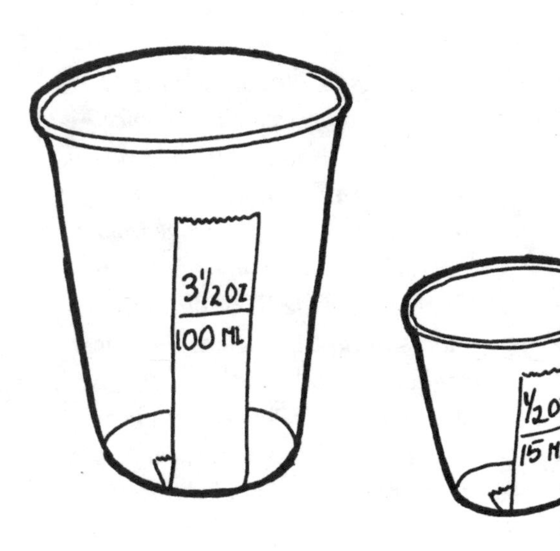

3. Using the graduated cylinder or measuring cup, make two measuring cups for each group. Measure out 15 ml (half an ounce) BTB in the measuring cup. Pour it into a small cup and then, with the tape and pen, note the depth on the side of the small cup. Repeat this procedure with a large cup for 100 ml (3½ oz.) of vinegar.

4. Read through the instructions for this session and practice:

   a. Generating carbon dioxide gas and collecting it in a balloon.

   b. Using the gas to put out a candle flame.

   c. Detecting carbon dioxide with BTB.

You may need to explain what baking soda is. Ask the class if anyone knows what it is used for at home? [cooking, to make carbon dioxide gas which causes cakes to rise; and cleaning, to neutralize the harmful acids produced by bacteria]

On the Day of the Activity

1. Provide each group of four students with half a cup of baking soda and two-thirds of a cup of vinegar.

2. For easy distribution, gather the equipment on a tray for each group of four students. Plan to use one set of the equipment for demonstration purposes.

3. Use a few drops of wax to secure the candle in the middle of the container for use during the demonstration. Place the wine bottle, vinegar, baking soda, a piece of paper, one graduated cylinder, and the candle and container near where you plan to conduct the demonstration of how to generate carbon dioxide gas.

The candle's wick should be below the rim of the glass.

Review

1. Ask the class to recall the results of the global warming game. You may wish to use questions such as the following to guide the review.

   - In Round 1, what percentage of the light photons from the Sun were absorbed by molecules and warmed the Earth?
     [about 50%]

   - In Round 2, you added carbon dioxide molecules to the atmosphere. What effect did these have on the fate of infrared photons?
     [They tended to stay around in the atmosphere longer, and were absorbed more frequently.]

   - What effect did increasing the number of carbon dioxide molecules in the atmosphere have in the game?
     [More infrared photons stayed around longer, and were absorbed even more frequently.]

2. Ask the students to take out their homework sheets, and discuss their answers to the questions about the relationship between global temperatures and carbon dioxide during the past 160,000 years.

   a. Over the past 160,000 years the temperature of the atmosphere has been as much as 18°F colder than it is now.

   b. Over the past 160,000 years the temperature of the atmosphere has been as much as 5°F warmer than it is now.

   c. It was at least as warm about 130,000 years ago as it is today.

   d. About 10°F.

   e. There was as much carbon dioxide in the air about 120,000 years ago as there was in 1850.

f. In the ice record from Antarctica, high temperatures corresponded with high concentrations of carbon dioxide. (Explain that when two sets of different data seem to respond in the same way, it is called a ***correlation***.)

g. 1995

h. Ask students for their opinions as to why temperature and carbon dioxide levels appear to be correlated.

3. Given your students' experiences in the previous sessions, they will probably conclude that the amount of carbon dioxide determines the average global temperature. Point out that scientists don't know for certain if changes in carbon dioxide caused changes in temperature, or if changes in temperature caused changes in the carbon dioxide levels. For example, it is possible that rising temperatures resulted in more forest fires, causing plants to release carbon dioxide, or $C0_2$ released by erupting volcanoes warmed the Earth, thus increasing the concentration of carbon dioxide in the atmosphere. The factors involved in analyzing world climate and temperature are complex.

4. Ask students what the **range in temperature** has been over the past 160,000 years.
   [About 23°F]
   Tell the class that temperature differences tend to be more extreme at the Poles and less extreme near the Equator. The Earth's average temperature today is about 59°F. During the Ice Ages, scientists believe that the average temperature of the atmosphere around the entire globe was about 9°F or 10°F colder than at present.

5. Remind the students that ever since the Earth has had an atmosphere, the greenhouse effect has warmed the Earth. However, over the past 100 years, the concentration of carbon dioxide has increased dramatically, and the average global temperature has increased by about 1°F. Many researchers suspect the increase in carbon dioxide has **caused** the increase in temperature. Although this is still a matter of scientific debate, the majority of researchers believe that, if we do not change the rate at which we burn fossil fuels, the Earth will become warmer by 2–7°F by the end of the next century.

Demonstration of Carbon Dioxide's Properties

1. Ask the class to tell you what they already know about carbon dioxide gas—what it looks like, where it is produced, and what are its properties.

2. Explain that to detect carbon dioxide the first thing you need is a good supply of the gas. Ask if anyone knows an easy way to make carbon dioxide.
   [Breathing is an easy way, but it doesn't produce pure carbon dioxide.]

3. Reveal that an easy way of making pure carbon dioxide is in a chemical reaction between two common substances: vinegar and baking soda. Ask the students to watch the demonstration because they will be doing it themselves.

4. Place an empty wine bottle in front of you. Measure 100 ml (3½ oz.) of vinegar and pour the vinegar into the wine bottle. Demonstrate how to make a funnel from a piece of paper to pour the baking soda powder into the bottle.

5. Ask for a volunteer from the class to come up and light the candle in the container, and then to stand nearby. Put four heaping teaspoons of baking soda into the bottle, using the funnel. Ask the volunteer to observe what is going on, and report this to the class. Ask the volunteer to place her finger over the end of the bottle for a few seconds and report what it feels like.

6. Ask the class to watch what happens, as you hold the opening of the bottle directly above the candle flame, and "pour" the gas **(not the liquid)** onto the flame.
   [The carbon dioxide will fill up the container, and block oxygen from getting to the candle, and so the flame will go out.]

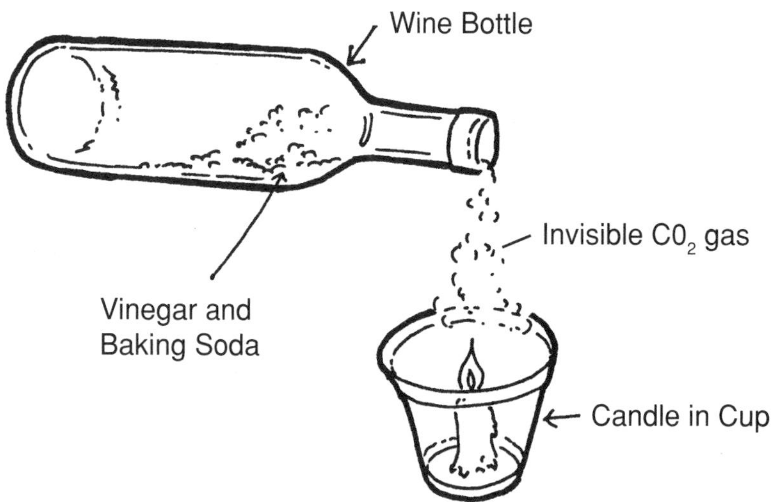

7. Rinse the bottle so it can be used by a group of students.

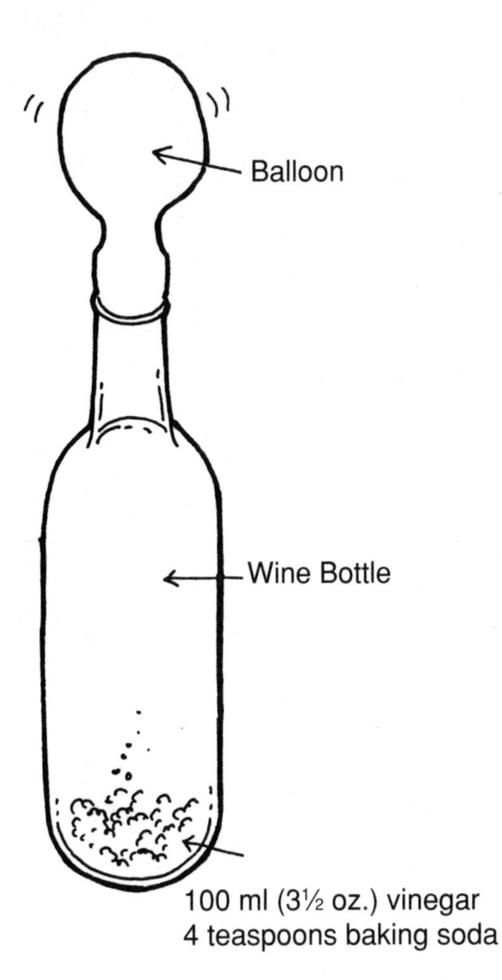

Collecting a Sample of Carbon Dioxide

1. Tell the students that the purpose of the laboratory activity they are about to conduct is to find out how to detect the presence of carbon dioxide. To do that they must first fill a balloon with carbon dioxide gas.

2. Tell them that they will prepare a sample of carbon dioxide gas as follows:

   a. Add 100 ml (3½ oz.) vinegar to a bottle, and then 4 heaping teaspoons of baking soda. Write on the board:

      100 ml (3½ oz.) vinegar
      4 heaping teaspoons baking soda

   b. The carbon dioxide will drive out all of the air in the bottle in less than a second. Quickly stretch the neck of a balloon over the opening to catch the escaping gas. The balloon should inflate to about 4 inches in diameter.

   c. Tightly tie off the neck of the balloon with a twist tie (helpful hint: twist the balloon neck immediately after collecting the gas, so no gas escapes before you get a chance to tie it off). Warn students not to puncture the balloon with the sharp end of the twist tie.

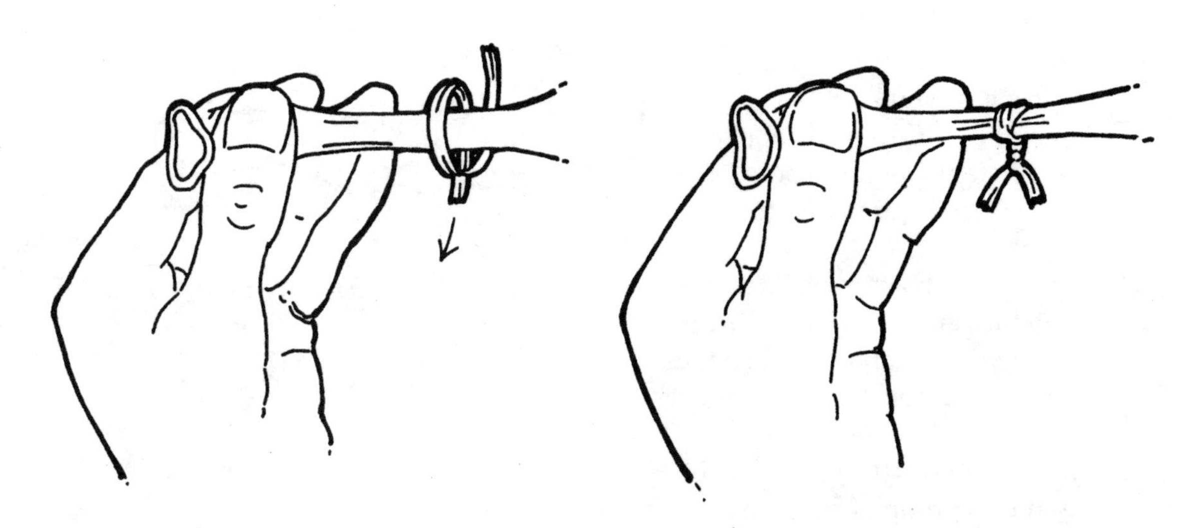

To seal the balloon, **wrap** the twist tie around the balloon twice, then twist the ends together.

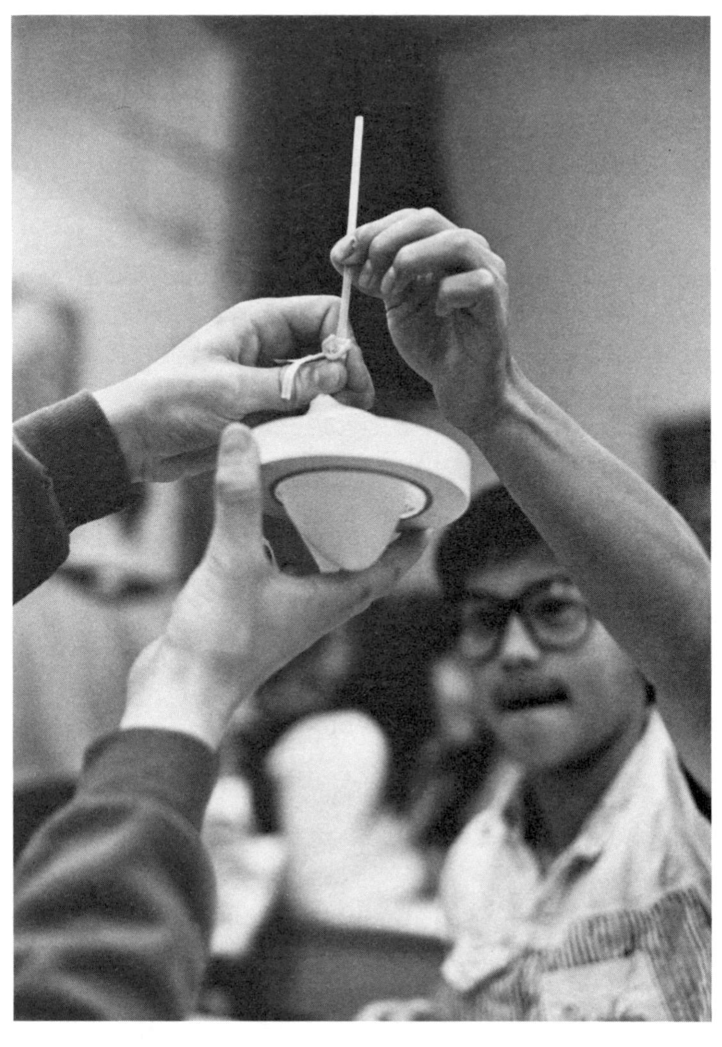

Use the inner diameter of a roll of masking tape to make the balloons the same size.

3. Tell the students they will have to work as a team, with one person holding the bottle, while someone else stretches the balloon over the end of the bottle.

4. Demonstrate how to use the air pump to fill a second balloon (of a different color) to approximately the same size as the balloon with carbon dioxide, with a sample of air from the room, and close it with a twist tie.

5. Have one student from each group pick up the materials on a tray. Pass the air pump to the first group. Allow the students to get to work and collect their gas samples.

6. Circulate around the room, helping as needed. If students lose their sample of carbon dioxide or puncture their balloon, have them add another 100 ml (3½ oz.) of vinegar to the contents of the bottle. There is enough baking soda in the bottle to generate carbon dioxide several times more.

Detecting Carbon Dioxide

While the students are working, prepare the next demonstration by measuring and pouring 15 ml (½ oz.) of the blue liquid BTB into a small clear plastic cup. Fill two more cups to the same level.

2. When the students have finished, demonstrate how to test for the presence of carbon dioxide with the assistance of a student volunteer.

We advise against bubbling the carbon dioxide gas through the BTB in the demonstration, as it will take away the surprise when the color changes from blue to yellow.

a. Insert the straw into the neck of the balloon with carbon dioxide, and seal it with a second twist tie.

b. Slowly loosen the top twist tie, releasing gas until the balloon will just pass through the hole in a roll of masking tape.

c. Clamp off the flow of gas with your fingers, and place the bottom end of the straw into the BTB.

d. Loosening your grip, slowly bubble the gas through the blue liquid. Squeeze the balloon to release all of the gas.

e. Observe the color of the liquid.

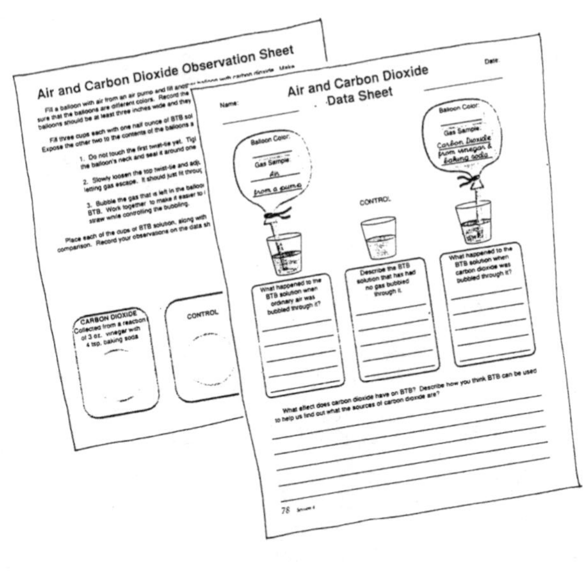
Air and Carbon Dioxide Observation Sheet

CARBON DIOXIDE

CONTROL

Air and Carbon Dioxide Data Sheet

CONTROL

3. Hold up an **Observation Sheet** (see page 74). Tell the students to put the Observation Sheet on the tray. Indicate where they should place their BTB samples.

4. Hold up a **Data Sheet** (see page 75), and explain how each individual is to record their results.

5. **Safety Reminder**: Point out that while none of these chemicals is dangerous, it is good to practice safe scientific methods. In this experiment leave all BTB samples on the trays in case they spill; don't splash any chemicals around; and, if available, wear safety goggles.

6. Ask if there are any further questions before starting. If not, have the students adjust the size of their balloons and bubble the gas samples through the BTB.

7. When all of the groups have finished, ask, "How does BTB react in the presence of almost pure carbon dioxide?" [It turns yellow.]

8. Tell the students that BTB stands for "bromothymol blue solution," which can be used to test for the presence of carbon dioxide gas. It changes color from blue to blue-green (a little carbon dioxide) to green (some carbon dioxide) to greenish-yellow (more carbon dioxide) to yellow (a lot of carbon dioxide). Draw this color scale on the chalkboard. This drawing will help the students realize that green is an intermediary color between blue and yellow, and that there are various other shades of color between blue and bright yellow.

If your students have studied acids and bases, as in the GEMS unit Of Cabbages and Chemistry, *you can explain that BTB is an acid-base indicator. The carbon dioxide gas reacts with the water to form carbonic acid. The carbonic acid turns BTB from blue to yellow. The more carbon dioxide there is in the sample, the more yellow BTB will turn.*

9. Ask the students to record their results on their data sheets, and then write a sentence or two to describe their results.

10. When they finish, ask them to put their equipment back on the trays and to return the trays to the preparation area.

Discussing Results and Assigning Homework

1. Ask the students to compare how the two samples of gas affected the BTB. Did the room air affect the BTB at all? If so, how many groups observed a color change?
[The results of the experiment depend on the exact concentration of BTB, and on the ventilation in the room. In most cases, the room air changes the color of BTB very little, if at all.]

2. Ask the students what this means in terms of the concentration of carbon dioxide in air.
[The amount of carbon dioxide is near the limit of the sensitivity of the BTB test. In fact, carbon dioxide makes up only 0.0369% of the atmosphere, according to 1995 measurements.]

3. Remind the class of the links between carbon dioxide levels in the atmosphere and the global temperature which they learned about in previous sessions. Ask if they know of some sources of carbon dioxide in the atmosphere. List these on the chalkboard or butcher paper.

4. Tell the students that in the next session they will use this method of testing for carbon dioxide to compare the amounts of carbon dioxide in two other gas samples: one from human breath and one from a car exhaust.

As in handing out previous homework assignments, you may need to ask questions that clarify what the graph in the article represents.

5. Hand out the two-sheet assignment "Surprise Increase in Atmosphere's CO_2." Ask the students to read the article and answer the questions in preparation for the next class.

6. Explain that the article refers to CO_2—which is the chemical formula for carbon dioxide. The formula means that one molecule of carbon dioxide (CO_2) is made from one atom of carbon (C) and two atoms of oxygen (O_2).

Contributions to Global Warming by Human Activity

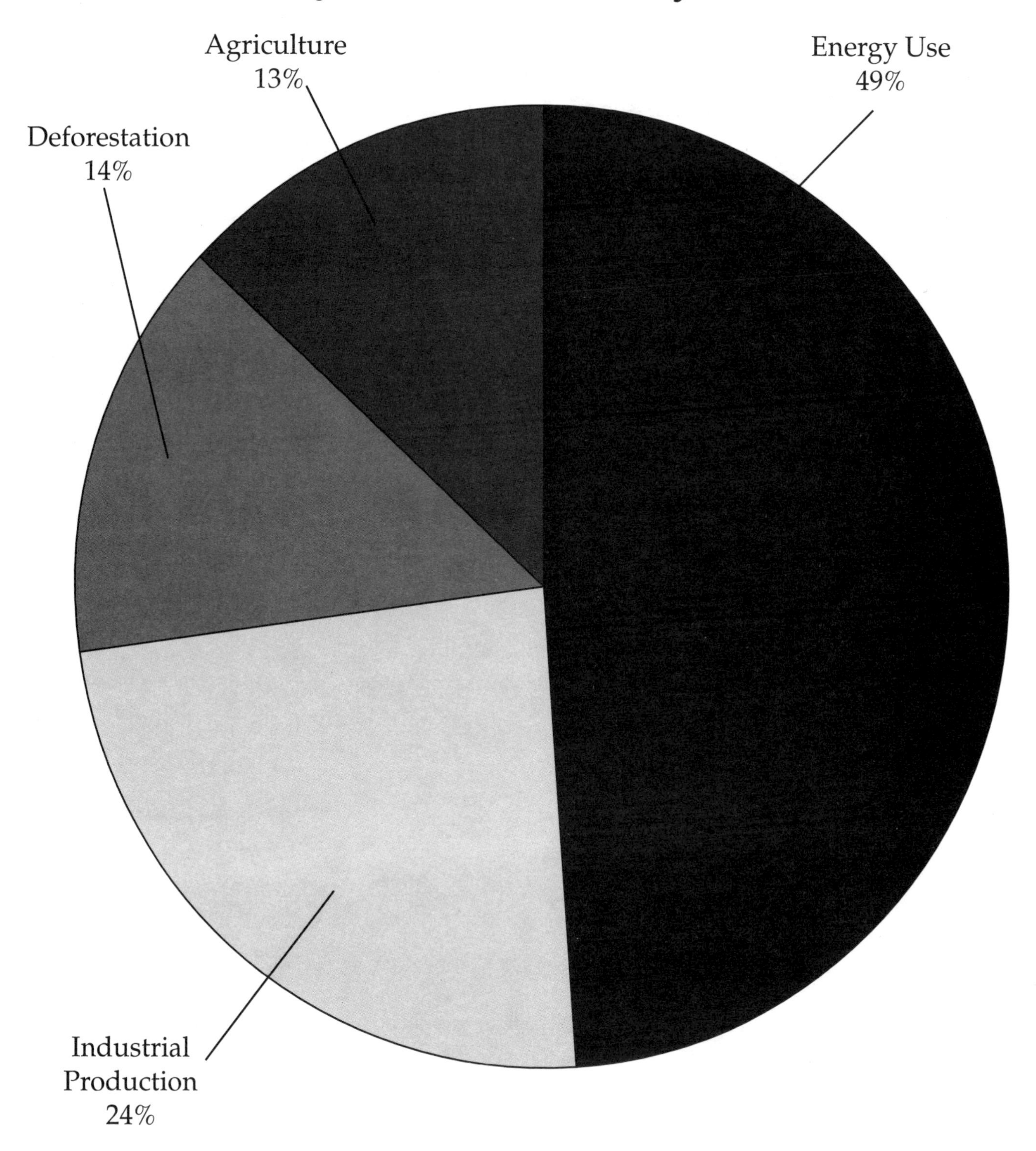

Source: World Resources Institute in collaboration with the United Nations Environmental Program, World Resources 1990–1991.

Name ____ Date ______________________

Surprise Increase in Atmosphere's CO_2!

The newspaper article on the next page describes a scientist's theory about why the amount of carbon dioxide is increasing in the atmosphere.

A. Notice that the graph of the amount of carbon dioxide jags up and down. How long does it take to go up and down once?

B. What does the article say caused the amount of carbon dioxide in the atmosphere to jag up and down?

C. What does the graph tell you about how the average concentration of carbon dioxide has changed from 1958 to 1988?

D. What is Dr. Keeling's opinion about why the concentration of carbon dioxide has increased?

E. Do scientists agree about how the climate will change as carbon dioxide increases? What opinions are expressed in the article?

F. Write at least one question about something you did not understand in the article, or something you would like to learn more about on this general topic.

Surprise Increase in Atmosphere's CO_2

By Charles Petit
Chronicle Science Writer

Air samples worldwide show a recent, mysterious surge in the rate of carbon dioxide increase in the atmosphere, a leading scientist is reporting.

The acceleration could mean that the greenhouse effect, and possible global warming from trapped sunlight, will occur faster than previously believed.

In the past year, the rate of increase may nearly have doubled, perhaps because of temporary climate changes. But a more troubling increase has been under way for at least 10 years, the new analysis suggests.

Although more burning of fossil fuel such as coal and oil is mainly to blame for the general rise in carbon dioxide, additional sources are blamed for 20 percent of the increase in the past decade, the measurements suggest.

These possible sources include soils and land plants reacting to changing atmosphere and temperature, particularly in arctic regions where a rise in temperatures may be speeding decomposition of vegetation residue.

Another source may be the cutting and burning of forests in the tropics and elsewhere.

San Diego Meteorologist

The assertions come from Charles David Keeling, a meteorologist at the Scripps Institution of Oceanography near San Diego. Keeling has kept the world's best records of the carbon dioxide increase since 1958 from a sampling station that he set up on Mauna Loa volcano in Hawaii. The station is now operated by the National Oceanic and Atmospheric Administration. His new study also includes data from stations from the arctic to the south pole.

"This recent evidence means we cannot rule out the possibility that a natural positive feedback could be accelerating the greenhouse effect over the short term," said Keeling and several colleagues in a paper prepared for testimony before the U.S. Senate Committee on Energy and Natural Resources. The hearing was to have been June 22 but was postponed. Keeling provided a copy of the paper to The Chronicle.

Keeling said in an interview that it could be that "CO_2 makes the greenhouse effect, and the greenhouse effect puts more CO_2 in the air" in a cycle that worsens the impact from burning fossil fuels alone.

If correct, the findings mean that work to slow the accumulation of greenhouse gases may have to involve much more than than just lowering use of fossil fuels, which release carbon dioxide when burned.

ATMOSPHERIC CARBON DIOXIDE

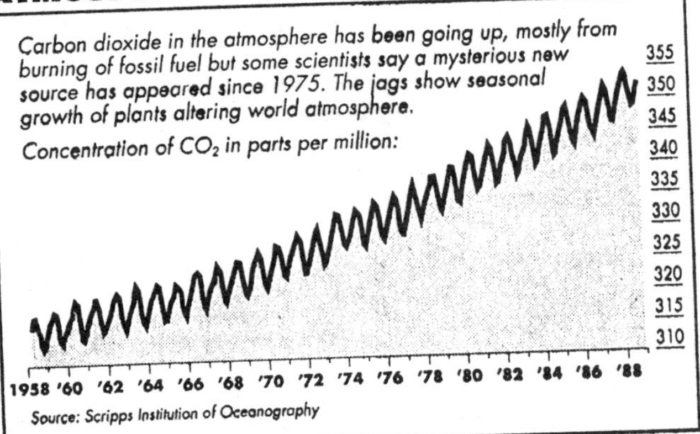

Carbon dioxide in the atmosphere has been going up, mostly from burning of fossil fuel but some scientists say a mysterious new source has appeared since 1975. The jags show seasonal growth of plants altering world atmosphere.
Concentration of CO_2 in parts per million:
355
350
345
340
335
330
325
320
315
310
1958 '60 '62 '64 '66 '68 '70 '72 '74 '76 '78 '80 '82 '84 '86 '88

Source: Scripps Institution of Oceanography

CHRONICLE GRAPHIC

25 Percent Increase

Since the middle of the last century, carbon dioxide has increased in the world's atmosphere by about 25 percent, going from around 280 parts per million to more than 350 parts per million now, and the amount may double in the next century. The main source is combustion products of coal, oil, gas, and other fossil fuels.

Rising levels of carbon dioxide lead many scientists to predict that the Earth is due for an unprecedented increase in average temperature.

The increase in carbon dioxide and other rising concentrations of "greenhouse" gases, including methane, nitrous oxides and chlorofluorocarbons, trap the heat from sunlight. Computer models suggest these gases will boost worldwide temperature three to eight degrees Fahrenheit by the middle to the end of the next century, levels higher than any in human history.

Numerous studies suggest that the effects may include dramatic shifts in weather and agriculture zones and a rise in sea level by one to three feet as glaciers melt and warmed seawater expands.

Exactly how the greenhouse effect may change climate has stymied scientists. Even the most powerful computers cannot chart the full complexities of a planet's weather. Some researchers say "negative feedbacks" could serve to damp out the greenhouse effect. A warming may create more clouds, for instance, which would tend to cool the surface and slow the rate of change.

Meeting on Pollution

Keeling is due to present his data today in Anaheim at a meeting of the Air and Waste Management Association, a trade group devoted to pollution issues.

Keeling's detailed new analysis of air samples suggests that since 1975, about nine billion tons of carbon, mainly in carbon dioxide, have been added to the atmosphere beyond what can be explained from fossil fuel burning.

Members of Keeling's group include Robert B. Bacastow, Justin Lancaster, and Timother P. Whorf of Scripps, which is part of the University of California, San Diego, and Willem G. Mook of the University of Groningen, the Netherlands.

Air and Carbon Dioxide Observation Sheet

Fill a balloon with air from an air pump and fill another balloon with carbon dioxide. Make sure that the balloons are different colors. Record the colors on your data sheet. The balloons should be at least three inches wide and they should be sealed tightly with twist ties.

Fill three cups each with one-half ounce of BTB solution. Set one aside as a control.

Expose the other two to the contents of the balloons as follows:

1. Do not touch the first twist-tie yet. Tightly tie another twist-tie over the balloon's neck and seal it around one end of a drinking straw.

2. Slowly loosen the top twist-tie and adjust the size of the balloon by letting gas escape. It should just fit through a ring three inches wide.

3. Bubble the gas that is left in the balloon into one of the cups of BTB. Work together to make it easier to hold the balloon and the straw while controlling the bubbling.

Place each of the cups of BTB solution, along with a control, in the spaces below for comparison. Record your observations on the data sheet.

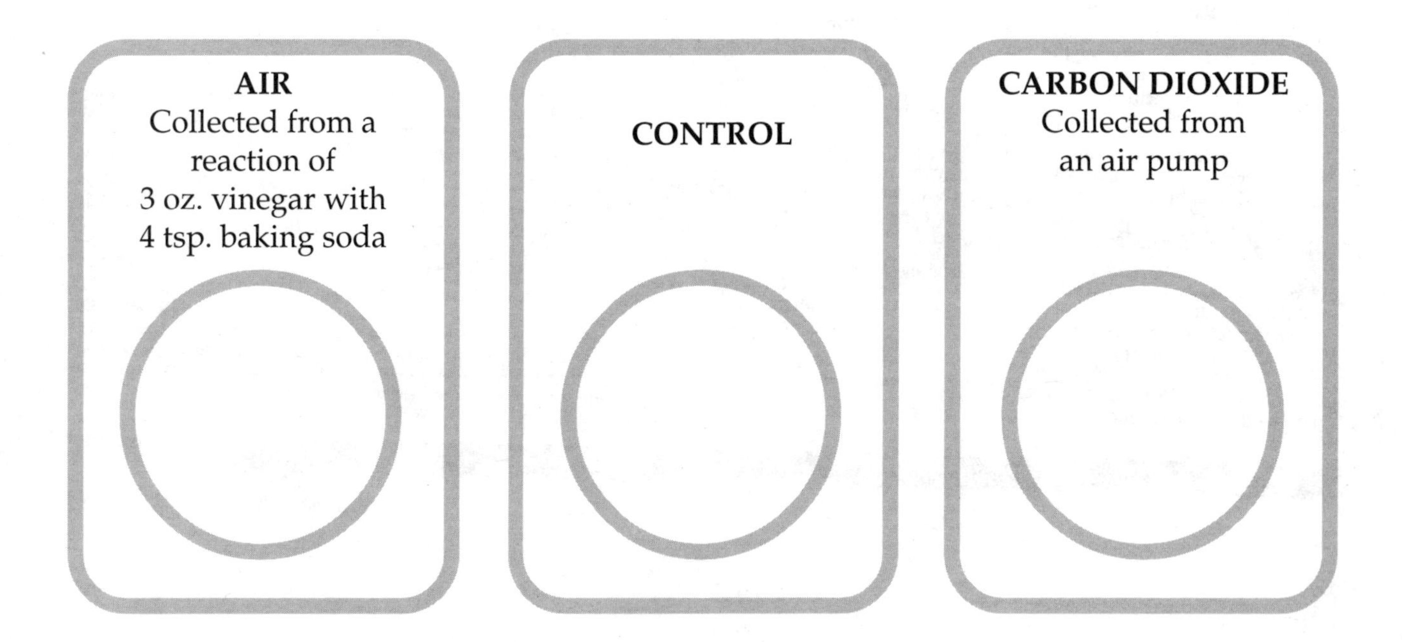

Name ____ Date ____________________

Air and Carbon Dioxide Data Sheet

Balloon Color

____

Gas Sample

Air

from a pump

Balloon Color

____

Gas Sample

Carbon Dioxide

from vinegar and baking soda

What happened to the BTB solution when ordinary air was bubbled through it?

____

____

____

____

____

Describe the BTB solution that has had no gas bubbled through it.

____

____

____

____

____

What happened to the BTB solution when carbon dioxide was bubbled through it?

____

____

____

____

____

What effect does carbon dioxide have on BTB? Describe how you think BTB can be used to find the sources of carbon dioxide?

____

____

____

____

____

Session 5: Sources of Carbon Dioxide in the Atmosphere

Overview

The common sources of carbon dioxide that contribute to the greenhouse effect are brought home to your students through this popular activity in which they compare the carbon dioxide content in their breath and in car exhaust.

In Session 4, your students practiced a technique for testing gas samples for carbon dioxide, using bromothymol blue (BTB). In this session they apply this technique in an experiment to investigate sources of carbon dioxide in their immediate environment. From these tests the students learn that the concentration of carbon dioxide exhaled by humans is less than in car exhaust. They discuss the implications of their results for controlling the amounts of carbon dioxide released into the atmosphere. Then, through a homework assignment, they learn about how much the various nations of the world contribute to the increase of carbon dioxide in the atmosphere.

The purposes of Session 5 are for the students to:

(1) improve their abilities to apply a chemical testing technique in conducting an experiment;

(2) practice sampling and testing procedures like those used in scientific studies of the atmosphere;

(3) draw out the distinction between "natural" and "industrial" sources of carbon dioxide in the atmosphere;

(4) communicate the relative contributions of the industrialized and developing nations to the global warming problem.

Do not be concerned about obtaining samples of gas from auto exhausts—it is really much easier than it might seem, and it's exciting for students to see the samples collected in front of their eyes! Observing how rapidly the balloons inflate is also a graphic demonstration of how much gas cars release into the atmosphere every second.

What You Need

For this session, you will need all of the materials prepared for Session 4 plus the items listed below. Items from Session 4 are marked with an asterisk.

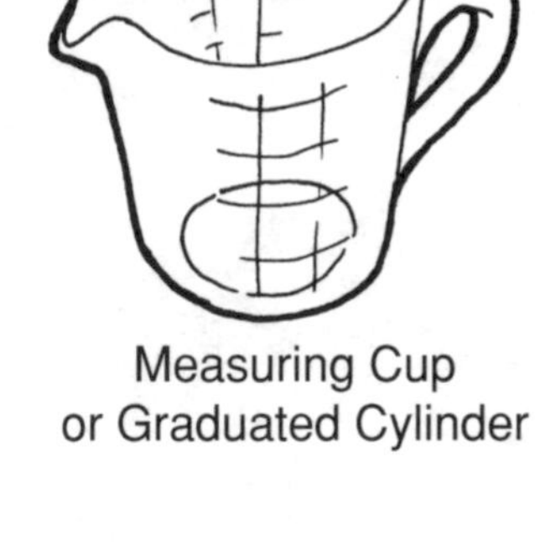

Measuring Cup
or Graduated Cylinder

For the class

- *❑ 1 box of baking soda
- *❑ 1 gallon of dilute BTB
- *❑ 1 air pump
- *❑ 1 gallon bottle white vinegar
- *❑ 1 graduated cylinder or measuring cup
- ❑ 1 car
- ❑ 1 manila folder
- ❑ 1 roll of masking tape or duct tape
- ❑ 1 chalkboard or butcher paper and marker

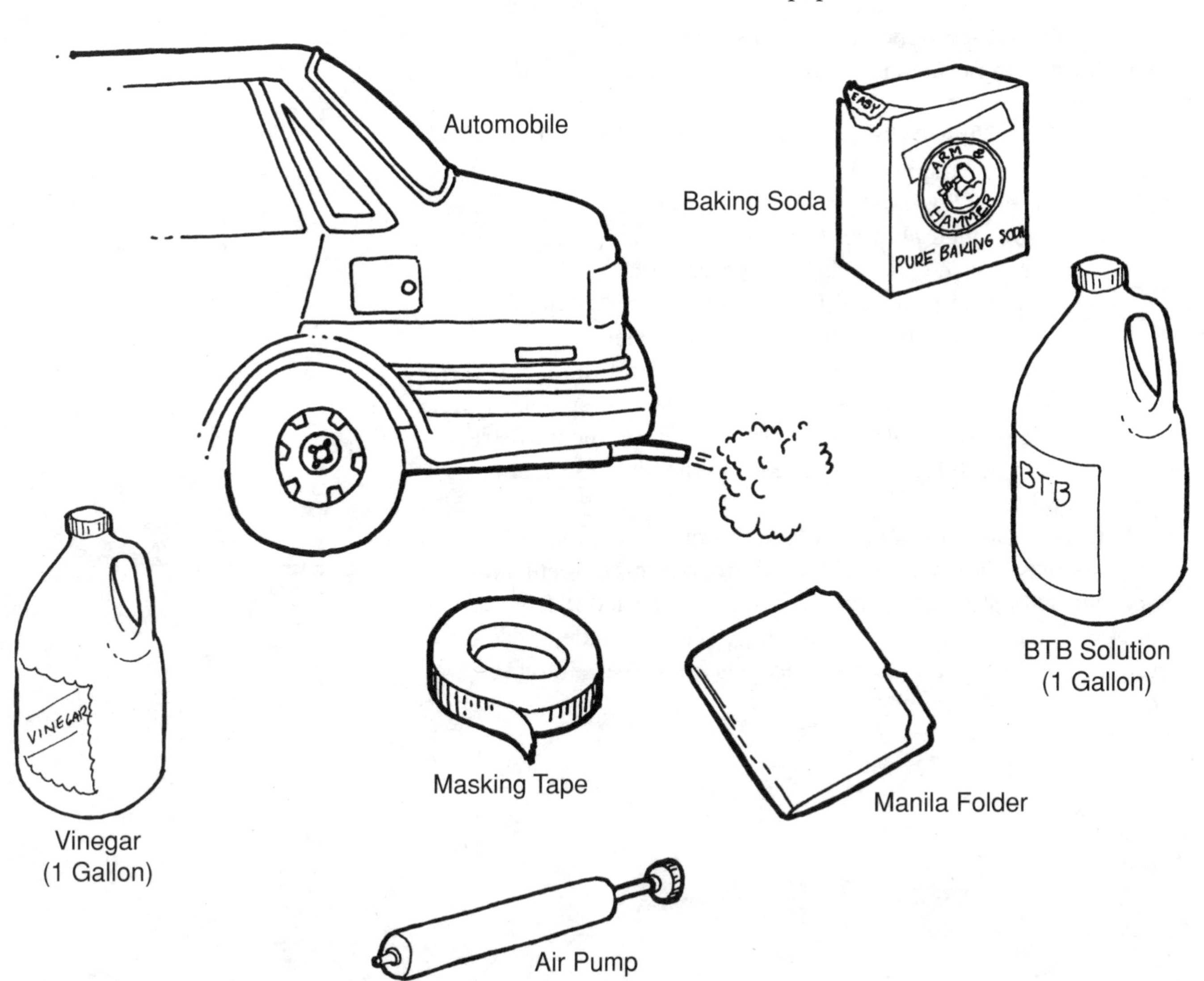

For each group of 4 students

- *❏ 1 empty glass wine bottle, 750 ml
- *❏ 1 dropper or squirt bottle containing BTB
- *❏ 1 teaspoon
- *❏ 3 plastic cups, 8 oz. size
- *❏ 5 small 3½ oz. clear plastic cups
- *❏ 4 balloons, 8–10" in diameter, three different colors
- *❏ 1 roll of masking tape or other object with a hole about 3½" in diameter; same size for all groups
- *❏ 1 tray
- ❏ 1 sheet of white paper
- ❏ 4 plastic straws
- ❏ 8 twist-ties
- ❏ 1 handout: "Four Gas Samples: Observation Sheet" (see page 88)

For each student

- ❏ 1 handout: "Four Gas Samples: Data Sheet" (see page 89)
- ❏ 1 two-sheet homework assignment: "Carbon Dioxide In the Atmosphere: Who Contributes and How Much?" (see pages 90–91)

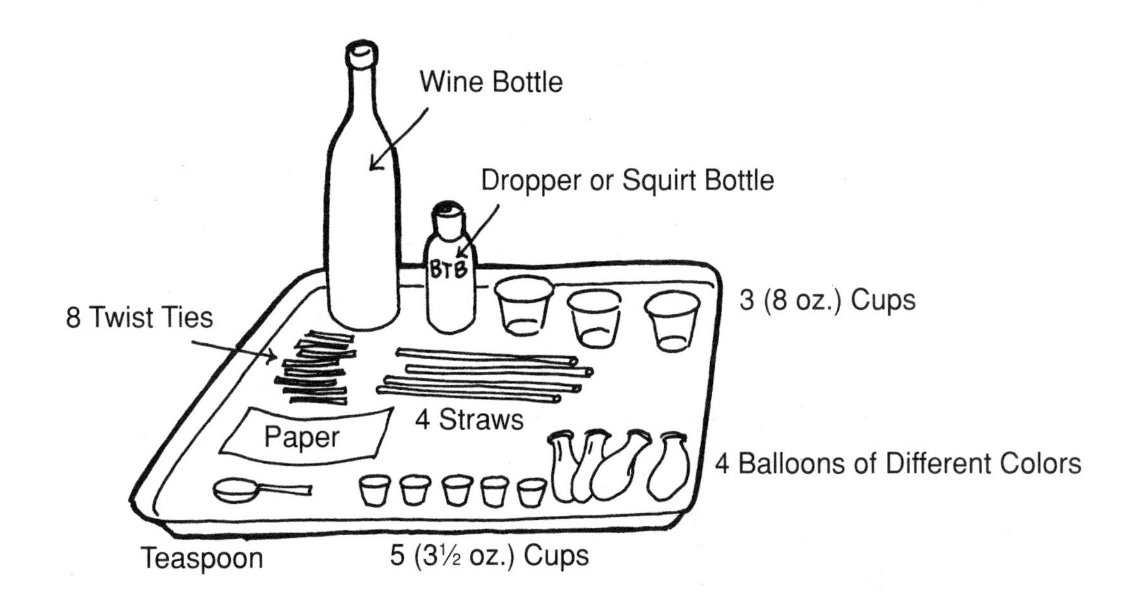

Getting Ready

Before the Day of the Activity

Opening must be large enough to fit over an exhaust pipe.

Opening must be small enough for balloon to fit over.

1. Use a car with a round exhaust pipe (square tail-pipes are difficult to seal). Prepare a cone for collecting car exhaust by rolling up a manila folder the long way. One end must be larger than the opening of a car's tail-pipe, and the other end must be small enough for a balloon to fit over it. Use plenty of tape to hold the cone in shape and to make the sides of the cone airtight. Trim the ends of the cone with scissors if necessary. Make a spare cone and have tape, folders, and scissors on hand when you collect the gas. Practice filling a balloon with car exhaust before class. Approach the exhaust pipe from the side and hold your breath when filling the balloons so you do not inhale the gases.

2. Make one copy of the two-sheet homework assignment: "Carbon Dioxide In the Atmosphere: Who Contributes and How Much?" and the data sheet for each student (see pages 89–91). Make one copy of the observation sheet for each group of four students (see page 88).

On the Day of the Activity

1. Park the car within close walking distance of your classroom. Preferably, the class should not have to cross a street to get to the car. Decide where the students should stand to watch you collect the gas samples so they can see what is happening, but are not in traffic. Since auto exhausts contain toxic gases, the students should stand back far enough so they do not have to breathe in car exhaust.

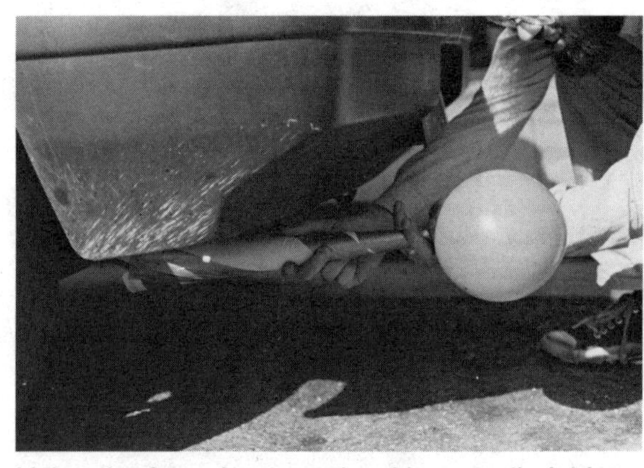

Using the funnel you made with a manila folder and tape, practice filling the balloon from the car exhaust pipe.

2. Assemble the equipment that each student team will need, as in Session 4, using trays for easy distribution. Use the amounts indicated above for this session.

3. Have the air pump available to be passed around from group to group.

Review

1. Discuss the students' answers to the homework questions (from worksheet on page 72 in Session 4):

   A. [The graph jags up and down thirty times over thirty years, or once per year.]

   B. [According to the caption on the graph, the jags are due to seasonal variations in plant growth.]

   C. [The average concentration of carbon dioxide has gone up from about 315 ppm to about 350 ppm.] (In 1995 it was about 360 ppm.)

   D. [According to Dr. Keeling, "more burning of fossil fuels, such as coal and oil, is mainly to blame for the general rise in carbon dioxide."]

   E. [Not all scientists agree about how the climate will change. Opinions include: more rapid warming than previously believed; or a slow down in warming due to increased cloud cover.]

   F. Ask for volunteers to state their questions, and ask for other students to respond. If questions cannot be answered, and you have time, write them on butcher paper, and add them to the class list.

2. Review the results from the previous session. [Blue BTB turned green, then yellow as carbon dioxide is bubbled through it; the level of carbon dioxide was not very high in air.] On the chalkboard, draw the scale from the previous session of how BTB changes color when exposed to different amounts of carbon dioxide.

ACIDS and BASES

The simplest atom is hydrogen. Its nucleus is a proton, which has a positive electrical charge. Around the nucleus is an electron, which has a negative electrical charge. When a hydrogen atom loses its electron, all that is left is the proton. A proton is symbolized as H+ since it is the nucleus of a hydrogen atom and has a net positive charge.

Chemicals made up of molecules from which protons (H+) may easily detach are called ***acids****. The protons attach themselves to other molecules floating in the water. These positively charged particles give the acid certain chemical properties. For example, when mixed with bromothymol blue (BTB), acids turn yellow.*

Other chemicals, called ***bases****, are composed of molecules from which pairs of hydrogen and oxygen atoms easily detach. Each of these pairs of atoms has a negative electrical charge, symbolized as OH-.*

When acids and bases are mixed in quantities that provide equal amounts of charged particles, the result is water: H+ added to OH- equals H_2O). This process is called neutralization.

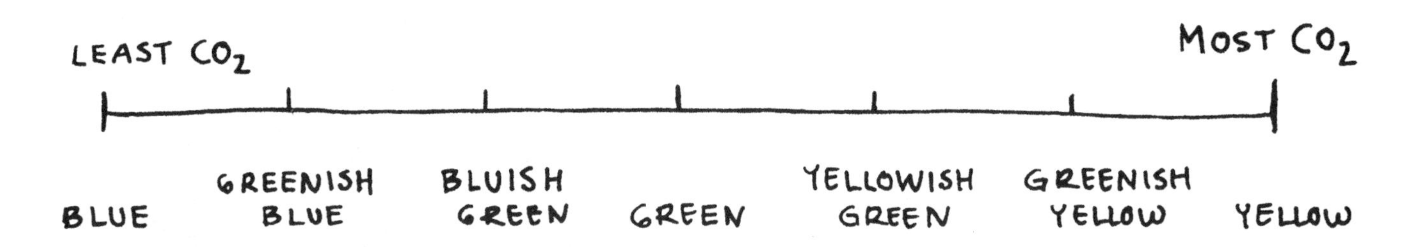

Collecting Samples of Car Exhaust

1. Explain to the class that among the sources of carbon dioxide in the atmosphere are car exhaust and the breath of humans and animals. The students will compare the concentration of carbon dioxide from a car's exhaust with other samples.

2. Tell the students they will take a brief field trip to a car to collect samples of exhaust, which they will analyze using the BTB method they practiced in the previous session.

3. Organize the class into groups, and issue each group one twist-tie, and one balloon. (Bring along several extra balloons and twist ties, in case you have some breakage. Also bring the funnel, as well as an extra funnel, folder, scissors, and tape).

4. Tell the students to stay in groups to follow you outside. When you get to the car, they will line up, and one group at a time will hand you their balloon for filling.

5. When you reach the car, tell the students that as soon as you fill their balloons, they are to work together to tie off the neck of the balloon tightly with the twist-tie. One student should hold the balloon closed, while the other student twists the neck of the balloon to prevent gas escaping and puts on the twist-tie.

6. Collect each of the samples in turn. Ask two volunteers to help you collect two extra samples in case a group loses its sample. Return to the classroom.

Conducting the Experiment

1. When the students are settled in their groups, tell them they will be testing four gas samples. List them on the board.

    1) Car exhaust
    2) Air from this room
    3) Carbon dioxide from vinegar and baking soda
    4) Human breath

2. Point out that the groups already have their car exhaust samples, and they learned how to collect the air sample in the last session by blowing up a balloon with an air pump.

3. Remind them they learned how to make and collect pure carbon dioxide by pouring 3½ oz. (100 ml) of vinegar into a large bottle, adding four level tablespoons of baking soda, and quickly pulling a balloon over the bottle opening. List the amounts on the chalkboard.

4. Ask the class how they think they should collect the human breath sample?
[By blowing up the balloon]

5. Tell the students that after they have collected all four gas samples, they should make them the same size by adjusting each balloon so it just passes through the hole in a large roll of tape—as in the previous session.

6. Hand out the data sheets and summarize the procedures.

   a. Pour half an ounce (15 ml) of BTB into each of five small plastic cups.

   b. Insert a plastic straw into the balloon's neck, and secure it with a twist tie.

   c. Loosen the top twist tie and adjust the balloon size to the diameter of the hole in the roll of tape, and clamp the neck between finger and thumb.

   d. With one person holding the straw in the cup, the other person carefully releases pressure on the balloon's neck, so all the gas slowly bubbles through the BTB solution.

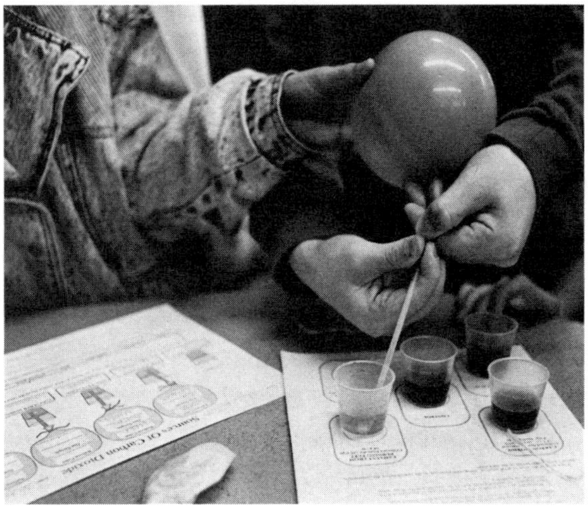

7. Tell the students to record on their data sheets which gas sample is put into which color balloon. After they do each test, they should record the color of the BTB in the space at the bottom.

8. Tell the groups to keep the BTB solution in the cups until the end of the experiment, so they can compare the colors. When they finish, they should fill in the names of the four samples along the line on their data sheets, from least to most carbon dioxide. Refer to the scale you have drawn on the chalkboard.

9. Remind the students to work together in pairs while doing the tests, and for everyone to take turns using the equipment.

10. Ask if there are any questions about the procedure. Then distribute the equipment trays, the bottles, and pass around the air pump so the students can get started.

11. During the activity, circulate among the groups to help as needed. Watch to see that no baking soda gets into any of the BTB tests (since baking soda is a base, it will neutralize carbonic acid, and prevent the BTB from changing to yellow/green). Also make sure students are viewing their BTB samples against the white paper of the observation sheet before they determine the color.

12. Remind students to fill in their results on the "color scale" on their data sheet, which indicates the relative amounts of CO_2 in the samples, before they discard any of their BTB test solutions.

13. As the students finish, tell them to fill in the final section of the data sheet, which asks them to express the findings in writing.

Some teachers suggest a good way to share class information at the end of Session 5 would be to prepare five clear cups, labeled as: Control, Exhaust, Breath, Air, CO_2. After students have filled in their data sheets, the teacher could circulate around the class, having students pour their samples into the appropriate cup. The colors in the cups would provide a color-based class average of the results. The clear cups would allow the colors to be seen on an overhead projector display as the class discusses the results.

Discussing the Results

1. When all groups have finished, have the class put their equipment (except for the data sheets) on the trays, and put the trays aside.

2. Ask the class which gas sample had the highest carbon dioxide content? Using the scale on the chalkboard, record the results. Fill in names of the other samples the same way. When there are conflicting results, list each gas sample and write the number of groups that had each result.

3. Ask the students to look at the data from all the lab groups and to say what they think the data tell us about the amount of carbon dioxide in equal volume samples of gas from these sources.
   [Most groups find that the order, from least to most carbon dioxide, is: air, human breath, auto exhaust, and almost pure carbon dioxide from the baking soda and vinegar reaction.]

4. Ask the class if the results confirmed their expectations or if they were surprised at the outcome.

5. Ask the students what additional information they would need to determine the relative amounts of carbon dioxide contributed to the atmosphere by humans breathing, and by humans driving cars.
   [They would need to know the number of humans versus the number of cars, how much time cars are driven on average, the volume per minute of gas exhaled by humans and "exhausted" by cars.]

6. Ask the class to suggest which of the sources of carbon dioxide they have measured (car exhaust and human breath) could be reduced? How might this be done?

More refined analyses of car exhausts show that these gases contain a high proportion of CO_2. In addition, there are other gases, including oxides of nitrogen (NOx) that react with water to form acids. These gases will contribute somewhat to the change in color of BTB from blue to yellow. Nitrous oxides are also greenhouse gases.

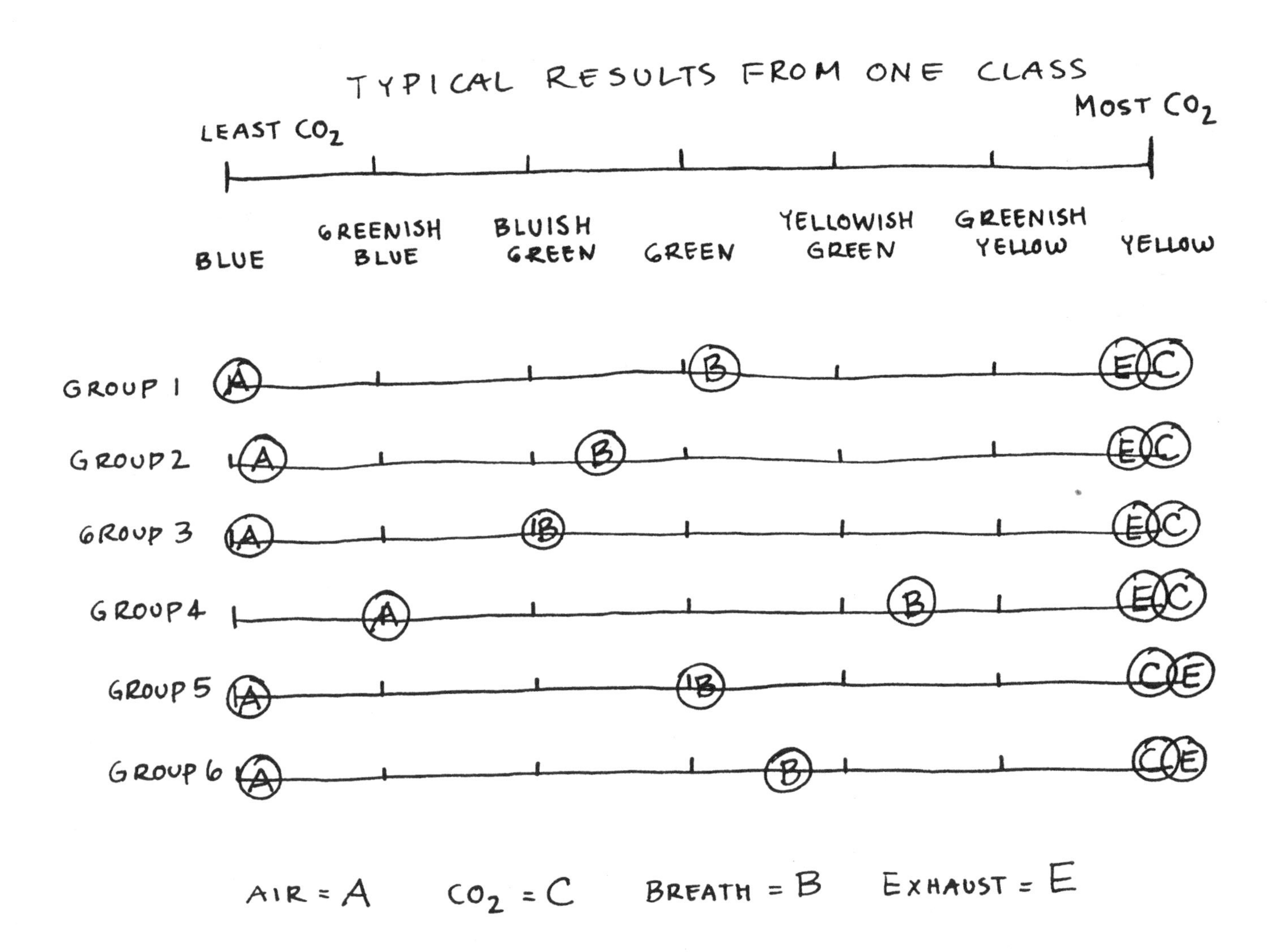

7. Remind the students that some carbon dioxide is needed in the atmosphere, or the Earth would freeze. The problem is one of balance. Many researchers think too much carbon dioxide is being added to the atmosphere from burning fossil fuels such as gasoline, natural gas, coal, and oil. These fuels are formed by the decay of plants and animals that lived long ago. **Each year people burn fossil fuels that required about 1 million years to form.** There is no quick way to recapture the carbon dioxide from the atmosphere and to replace those fuels. Scientists who have measured the amount of carbon dioxide from different sources found carbon dioxide released by the United States comes from the following sources:

   Cars and other transportation 8%

   Homes and businesses.................................. 15%

   Industry 28%

   Electric power plants 49%

   (*Note:* Electric power plants provide energy to homes and industries.)

   Source: World Resources 1993–94, *World Resources Institute, Oxford University Press, New York, N.Y. 1994. p.18.*

8. Hand out the homework assignment, "Carbon Dioxide In the Atmosphere: Who Contributes? And How Much?" Discuss the bar graphs to make sure the students understand what they represent. For example, ask for a volunteer to tell you:
   - "What is the current population of China?"
   - "What percentage of the world's population lives in the United States?"
   - "What percentage of the world's carbon dioxide is produced by the Russian Federation?"

   Tell the students to write answers to the questions by the next session.

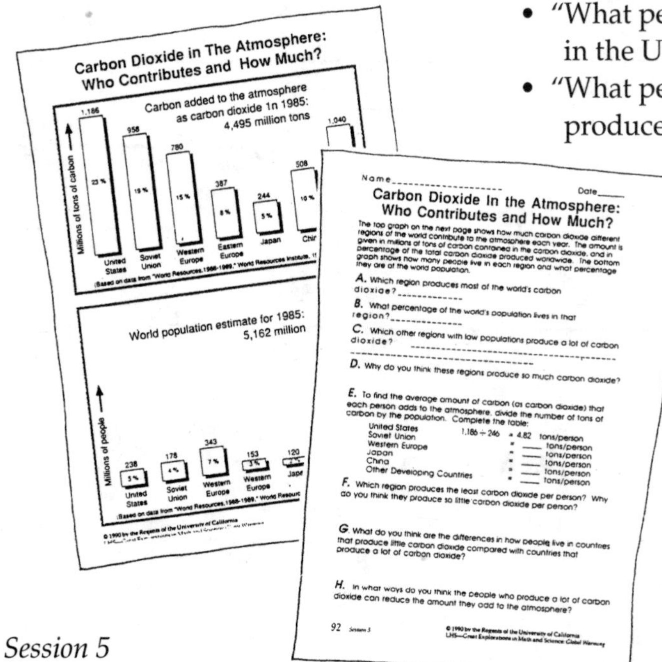

Name__________________ Date_____

Carbon Dioxide In the Atmosphere: Who Contributes and How Much?

The top graph on the next page shows how much carbon dioxide different regions of the world contribute to the atmosphere each year. The amount is given in millions of tons of carbon contained in the carbon dioxide, and in percentage of the total carbon dioxide produced worldwide. The bottom graph shows how many people live in each region and what percentage they are of the world population.

A. Which region produces most of the world's carbon dioxide? ____________

B. What percentage of the world's population lives in that region? ____________

C. Which other regions with low populations produce a lot of carbon dioxide? ____________

D. Why do you think these regions produce so much carbon dioxide?

E. To find the average amount of carbon (as carbon dioxide) that each person adds to the atmosphere, divide the number of tons of carbon by the population. Complete the table:

United States 1,186 ÷ 246 = 4.82 tons/person
Soviet Union = ____ tons/person
Western Europe = ____ tons/person
Japan = ____ tons/person
China = ____ tons/person
Other Developing Countries = ____ tons/person

F. Which region produces the least carbon dioxide per person? Why do you think they produce so little carbon dioxide per person?

G. What do you think are the differences in how people live in countries that produce little carbon dioxide compared with countries that produce a lot of carbon dioxide?

H. In what ways do you think the people who produce a lot of carbon dioxide can reduce the amount they add to the atmosphere?

92 Session 5

© 1990 by the Regents of the University of California
LHS—Great Explorations in Math and Science: Global Warming

Going Further

1. Here are some questions for further investigation that have particular relevance to global warming issues. Any or all of these questions could be investigated by students, by collecting a sample of gas in a balloon from each source, adjusting the balloon to a standard size, and testing it with BTB, as described in this session.

   - Do older cars produce a higher concentration of carbon dioxide in their exhaust than newer cars?

   - Do cars with pollution control equipment produce less carbon dioxide than cars that do not have such equipment?

   - Do diesel engines produce more or less carbon dioxide than gasoline engines?

   - How do various types of internal combustion engines differ in their carbon dioxide output:

     two-stroke vs. four stroke
     rotary vs. piston
     four cylinder vs. six cylinder
     car vs. motorcycle

2. Have students use a plastic garbage bag to collect a volume of gas exhausted by a car in a certain time period. Have them record the time it takes to fill the bag with that volume, estimate the volume of gas collected, and the amount of time an average car would run each day. From that data, the total estimated amount of exhaust gas that a car generates in a year can be calculated.

   Students could go on to estimate the volume of human exhaust (breath!) exhaled by one person in a year. Ask them to figure out how they might go about estimating this.

   Finally, students could compare their estimates of the volume of exhaust contributed by a car in one year to that contributed by one person in one year.

   The reasoning used and conclusions reached could provide the basis for an excellent discussion of the significance of these estimates when considering possible solutions.

Four Gas Samples: Observation Sheet

Collect all four samples of gas before making any tests. Each sample should be sealed in a balloon with a twist-tie. Make sure each balloon is a different color and each balloon is at least three inches wide. Record the colors of the balloons and their contents on your data sheet.

Test each of the samples for carbon dioxide as follows:

1. Do not touch the first twist-tie yet. Tightly tie another twist-tie over the balloon's neck and seal it around one end of a drinking straw.

2. Slowly loosen the top twist-tie and adjust the size of the balloon by letting gas escape. It should just fit through a ring three inches wide.

3. Bubble the gas that is left in the balloon into one of the cups of BTB. Work together to make it easier to hold the balloon and the straw while controlling the bubbling.

CARBON DIOXIDE
Collected from a reaction of 3 oz. vinegar with 4 tsp. baking soda

Place each of the cups of BTB solution, along with a control, in the spaces below for comparison. Record your observations on the data sheet.

EXHAUST FROM BURNING FUEL
Collected from the tail pipe of a car

CONTROL

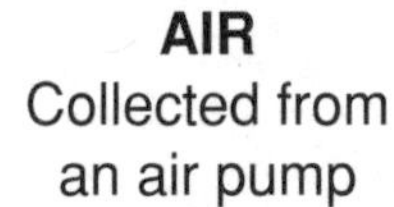

AIR
Collected from an air pump

HUMAN BREATH
Collected from a student's lungs

Four Gas Samples: Data Sheet

Name:____________________ Date:____________________

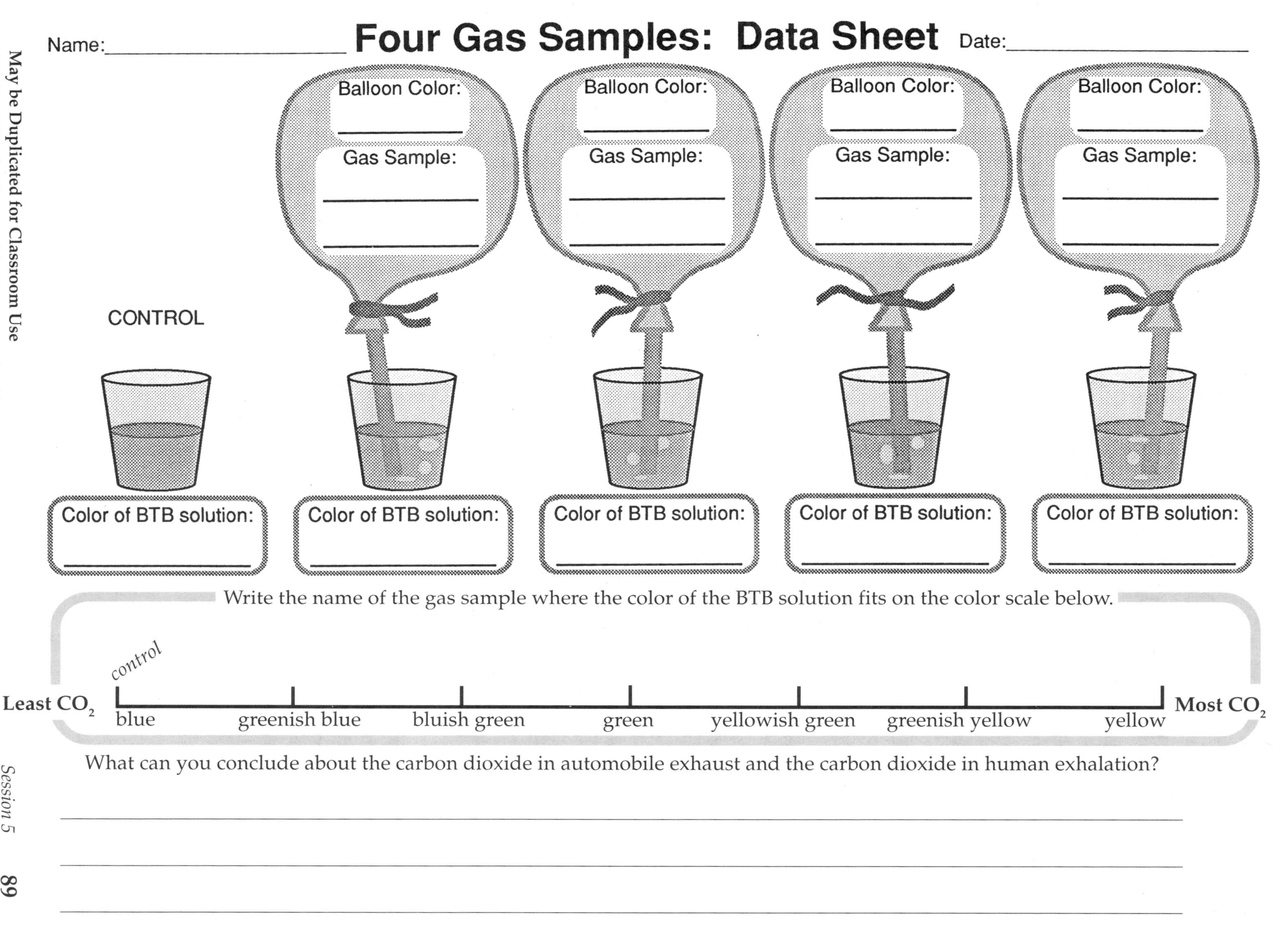

What can you conclude about the carbon dioxide in automobile exhaust and the carbon dioxide in human exhalation?

____

____

____

Name ______________________________ Date ______________

Carbon Dioxide In the Atmosphere: Who Contributes and How Much?

The top graph on the next page shows how much carbon dioxide different countries of the world contribute to the atmosphere each year. The amount is given in millions of tons of carbon contained in the carbon dioxide, and in percentage of the total carbon dioxide produced worldwide. The bottom graph shows how many people live in each country and what percentage they are of the world population.

A. Which country produces most of the world's carbon dioxide? ______________

B. What percentage of the world's population lives in that country? ______________

C. Which other countries with low populations produce a lot of carbon dioxide? ______________________

D. Why do you think these countries produce so much carbon dioxide?

E. To find the average amount of carbon (as carbon dioxide) that each person in each country adds to the atmosphere, divide the number of tons of carbon by the population. Complete the table.

United States	4,881 ÷ 256	=	19	tons/person
China		=	______	tons/person
Russian Federation		=	______	tons/person
Japan		=	______	tons/person
Germany		=	______	tons/person
India		=	______	tons/person
All Other Countries		=	______	tons/person

F. Which country produces the least carbon dioxide per person? Why do you think they produce so little carbon dioxide per person?

G. What do you think are the differences in how people live in countries that produce little carbon dioxide compared with countries that produce a lot of carbon dioxide?

H. In what ways do you think the people who produce a lot of carbon dioxide can reduce the amount they add to the atmosphere?

Carbon Dioxide in the Atmosphere: Who Contributes and How Much?

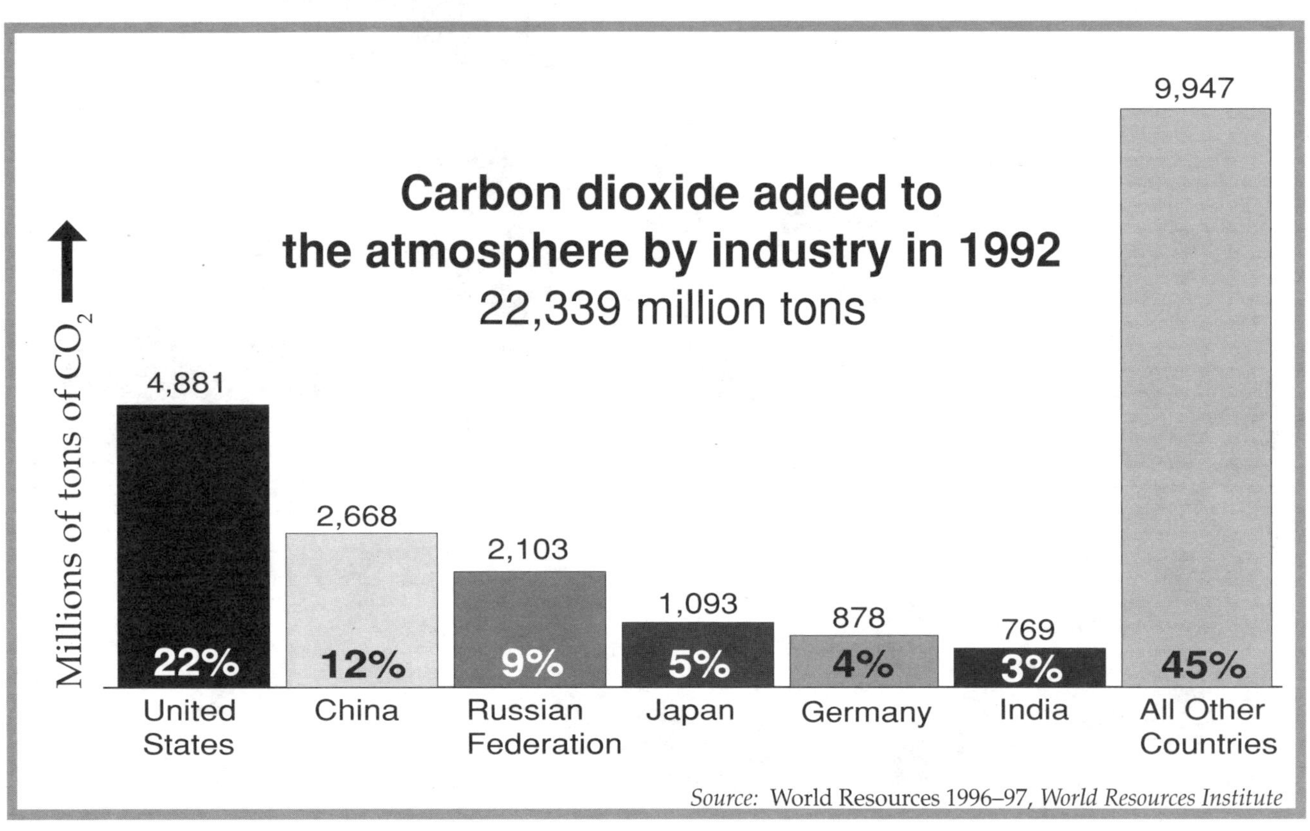

Source: World Resources 1996–97, *World Resources Institute*

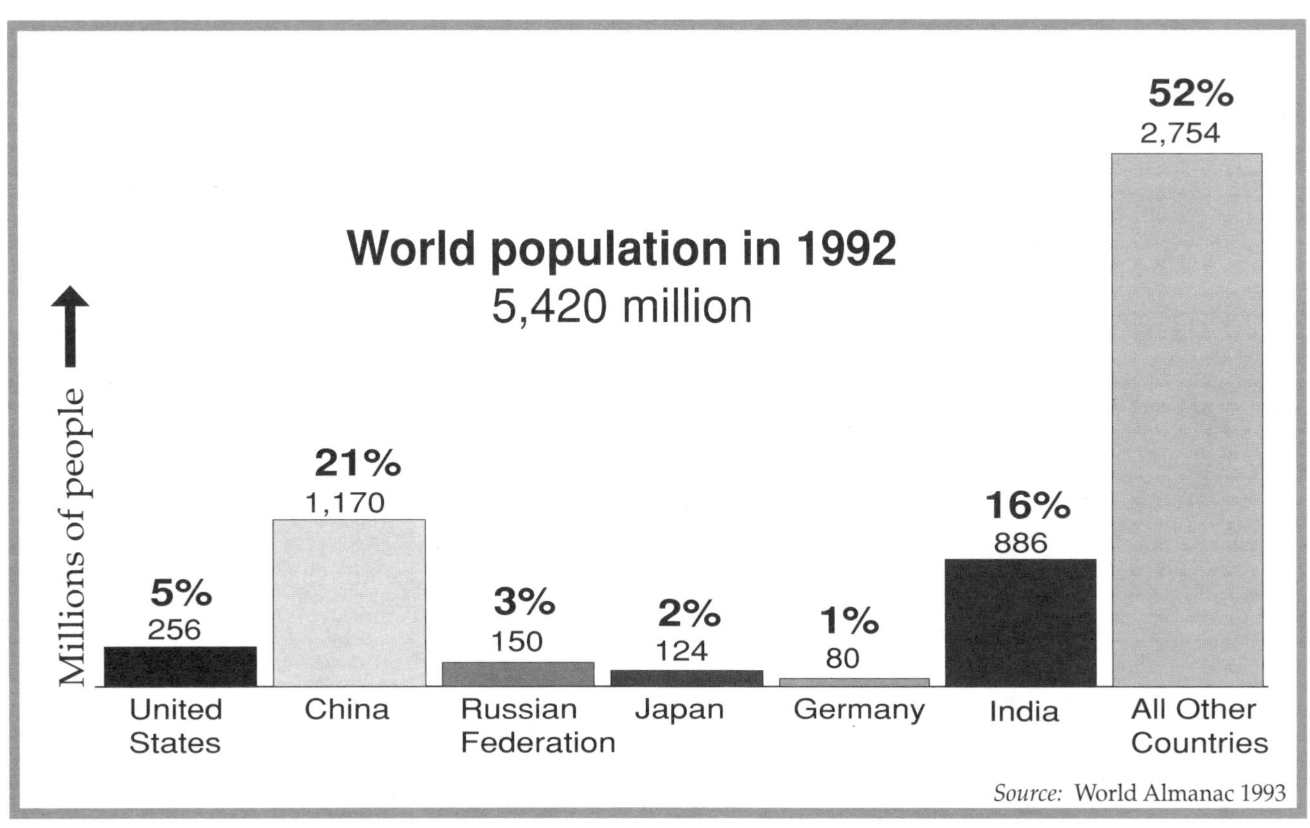

Source: World Almanac 1993

CAR MANUFACTURERS

Session 6: Changes On Noua's Island

Overview

So far in this guide, emphasis has been on understanding how the greenhouse effect works, and its consequences for global warming. Now the focus shifts to the large-scale effects of global warming, and some of the issues and dilemmas our global society faces in dealing with these effects.

Session 6 consists of two parts. First, the students meet in small groups to plan how they will represent specific points of view at a mock "World Conference on Global Warming" (to be conducted in Session 8). Summaries of these plans allow the teacher to provide feedback to the groups when they refine their plans during the next session. In part two, the students hear an imaginative story ("Noua's Island"), which helps them visualize the possible effects global warming will have on the lives of people in a distant place on the Earth.

The purposes of Session 6 are to:

(1) introduce the concept of an international forum for discussing environmental problems and possible solutions;

(2) provide students with insights into the effects of global warming on a society and culture very different from their own;

(3) provide the experience of looking at an environmental issue from viewpoints that may be different from their own.

You may find your class needs more time to discuss the homework, "Noua's Story," and other ideas covered so far in the unit. If so, it is probably best to add another session rather than rushing your students to stay on schedule. Most of Session 6 could then be taken up by discussion of the homework and Noua's Story, with the World Conference introduced at the end. This arrangement would also give students time to think about their preferred group until the next class period. Then, the students would form their interest groups and start planning for the World Conference.

What You Need

For each student interest group

❏ One copy (for each student in the group) of a specific interest group's description and list of questions. Include one additional copy for the group (see pages 101–105)

For each student

❏ Homework article on global climate change (select the article most appropriate for your students—see pages 106–107)

Getting Ready

1. Photocopy pages 101–107 for in-class and homework reading. Depending on your students, you may wish to use one or both of the articles.

2. In this session, students form interest groups to prepare for a "World Conference on Global Warming." Decide in advance if it is best for you to assign students to the five groups on the basis of which students work productively together, or if you want them to choose which of the groups they wish to represent.

3. Plan how you will organize tables or desks so small groups can meet together halfway through the session.

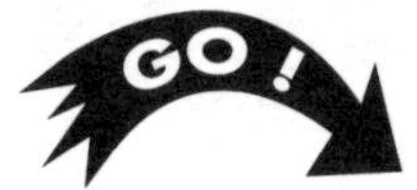

Where Does the World's CO_2 Come From?

1. Discuss the homework questions from Session 5 (see page 90).

   A. [The United States produces most of the world's carbon dioxide.]

   B. [Five percent of the world's population lives in the United States.]

   C. [The Russian Federation, Japan, and Germany have low populations and produce a lot of carbon dioxide.]

   D. [Some countries produce more carbon dioxide per person because they have more industry and more automobiles than other countries.]

   E. [United States .... 19.06 tons/person
   China .... 2.28 tons/person
   Russian Federation .... 14.02 tons/person
   Japan .... 8.81 tons/person
   Germany .... 10.97 tons/person
   India .... .86 tons/person
   All Other Countries .... 3.61 tons/person]

F. [India. A lower level of industrialization, fewer automobiles, and large populations in many developing nations probably account for the lower production of carbon dioxide per person.]

G. Allow individuals to express their opinions and discuss their different points of view on this question.

H. Allow individuals to express their opinions and discuss their different points of view on this question.

The Story of Noua's Island

1. Ask students to imagine themselves in a developing country that may be affected by increased sea levels. Although people in that country contributed very little to the cause of global warming—increased carbon dioxide—they will be affected by it as much as any of the industrialized nations, and in some cases the effects will be even greater.

2. Tell the class you are about to read a story to help them imagine what it is like to live in a developing country, and how people in that country may be directly affected by global warming. Explain that the story is not about an actual person, but it does describe a real situation for millions of people who live on islands or low coastal areas around the world. The story portrays predictions by researchers concerning how people in these areas are likely to be affected by rising sea levels if global temperatures increase by just a few degrees Fahrenheit.

3. Ask the class to put aside any distractions, sit back, relax, and imagine what is happening in the story.

4. Begin to read the story **slowly and carefully**—it is meant to transport the students into the world of an ocean island faced with the problem of rising sea levels. (Your reading copy of the story is on pages 98–100.) Where it is indicated in the story, stop and ask questions.

5. After reading the story, ask the students: What would you do if you were in Noua's situation? Do you think other countries are going to respond to this dilemma of the low-lying ocean islands? Why or why not? Allow the students to discuss their opinions, and explain they will have a greater opportunity to plan what might be done during an upcoming conference on global warming.

Planning the International Conference

1. Tell the students they will conduct an imaginary **World Conference** to discuss how people might cooperate to decrease the greenhouse effect, or cope with the changes it causes. In the conference, groups of students will represent people from various countries and other interests.

2. Divide the class into five groups (or allow the students to choose the interest group they will represent) as follows:
   Automobile Manufacturers
   Island Nations
   Agriculturists
   Conservationists
   Wood and Paper Producers

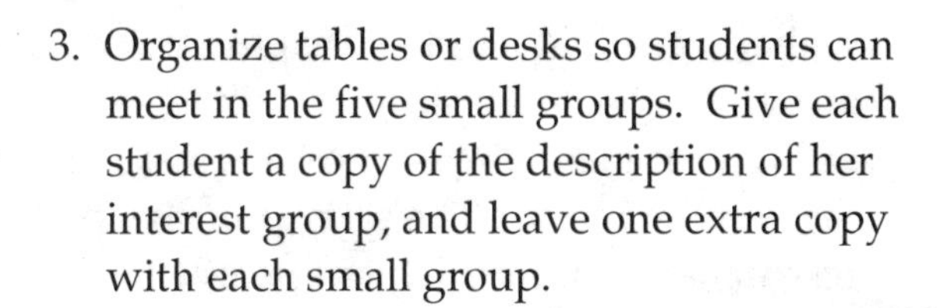

3. Organize tables or desks so students can meet in the five small groups. Give each student a copy of the description of her interest group, and leave one extra copy with each small group.

4. Briefly describe each of the interest groups for the entire class.

5. Explain that the World Conference has been called to allow each of the interest groups involved to present their point of view, and to look for possible solutions that are acceptable to some or all of the other interest groups.

6. Emphasize that it will not be a debate—everyone is affected by global warming, so no one is really for or against it. Instead, the task is to find out who is affected, how, and what, if anything, can be done about it.

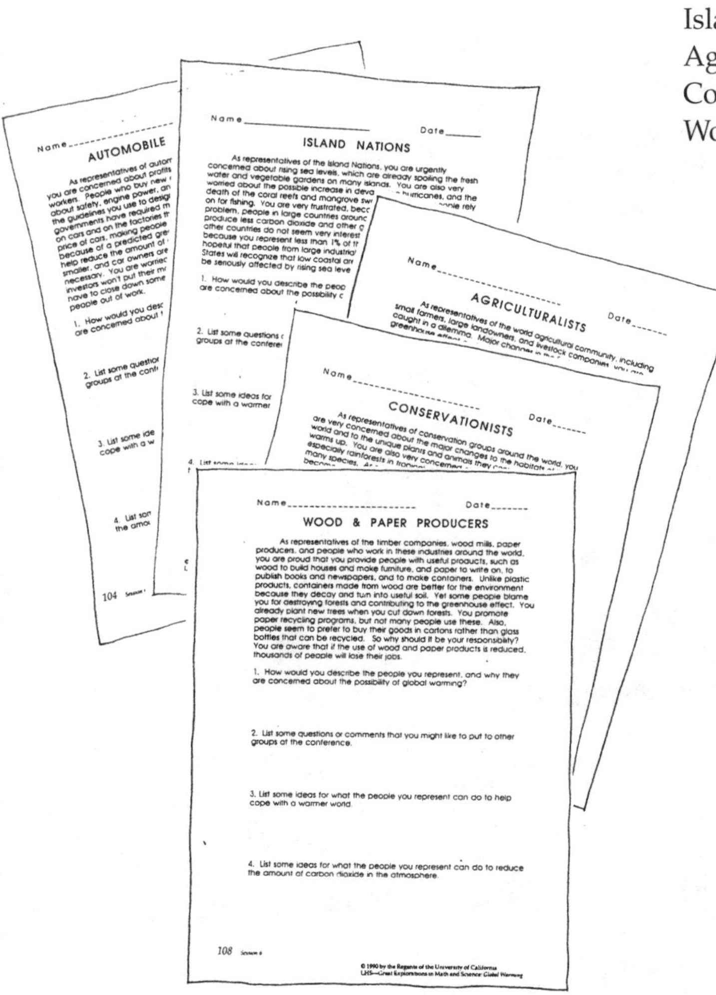

Name______________________ Date_______

WOOD & PAPER PRODUCERS

As representatives of the timber companies, wood mills, paper producers, and people who work in these industries around the world, you are proud that you provide people with useful products, such as wood to build houses and make furniture, and paper to write on, to publish books and newspapers, and to make containers. Unlike plastic products, containers made from wood are better for the environment because they decay and turn into useful soil. Yet some people blame you for destroying forests and contributing to the greenhouse effect. You already plant new trees when you cut down forests. You promote paper recycling programs, but not many people use these. Also, people seem to prefer to buy their goods in cartons rather than glass bottles that can be recycled. So why should it be your responsibility? You are aware that if the use of wood and paper products is reduced, thousands of people will lose their jobs.

1. How would you describe the people you represent, and why they are concerned about the possibility of global warming?

2. List some questions or comments that you might like to put to other groups at the conference.

3. List some ideas for what the people you represent can do to help cope with a warmer world.

4. List some ideas for what the people you represent can do to reduce the amount of carbon dioxide in the atmosphere.

108 Session 6

7. Tell the groups to appoint a Reader, who will read the interest group description and questions aloud to the group, and a Recorder to write down everyone's ideas in response to the four questions on the page. Allow the groups 10–20 minutes to read the handouts, and to discuss and record their ideas.

8. Five minutes before the end of the session, ask if there are any questions about the conference. Tell the class they will have some more time in the next session to plan what they are going to say at the conference.

9. Collect the writing from each of the groups so you can provide feedback during the next planning period in Session 7.

10. Hand out the reading assignment (pages 106–107). Instruct the class to read these before the next session, because they will find the information helpful in their planning for the "World Conference on Global Warming."

Going Further

1. Have students draw pictures of what they imagine Noua's Island to be like—an island nation threatened by rising sea levels due to global warming.

2. Students could write a letter to Noua, either for homework, or as a summary at the end of the unit, explaining to the island people what is currently known about global warming, and what solutions there might be to the problems that it may cause.

3. The class could write a play about Noua and the issue of global warming, to present to other classes, or dramatize an imaginary interview with Noua, asking what effects global warming may have on the life-style of the island nations. (Some helpful references are listed in the Resources, starting on page 146.) You may have other ideas about a play or other literary extension focusing on the topic of global warming.

Noua's Island

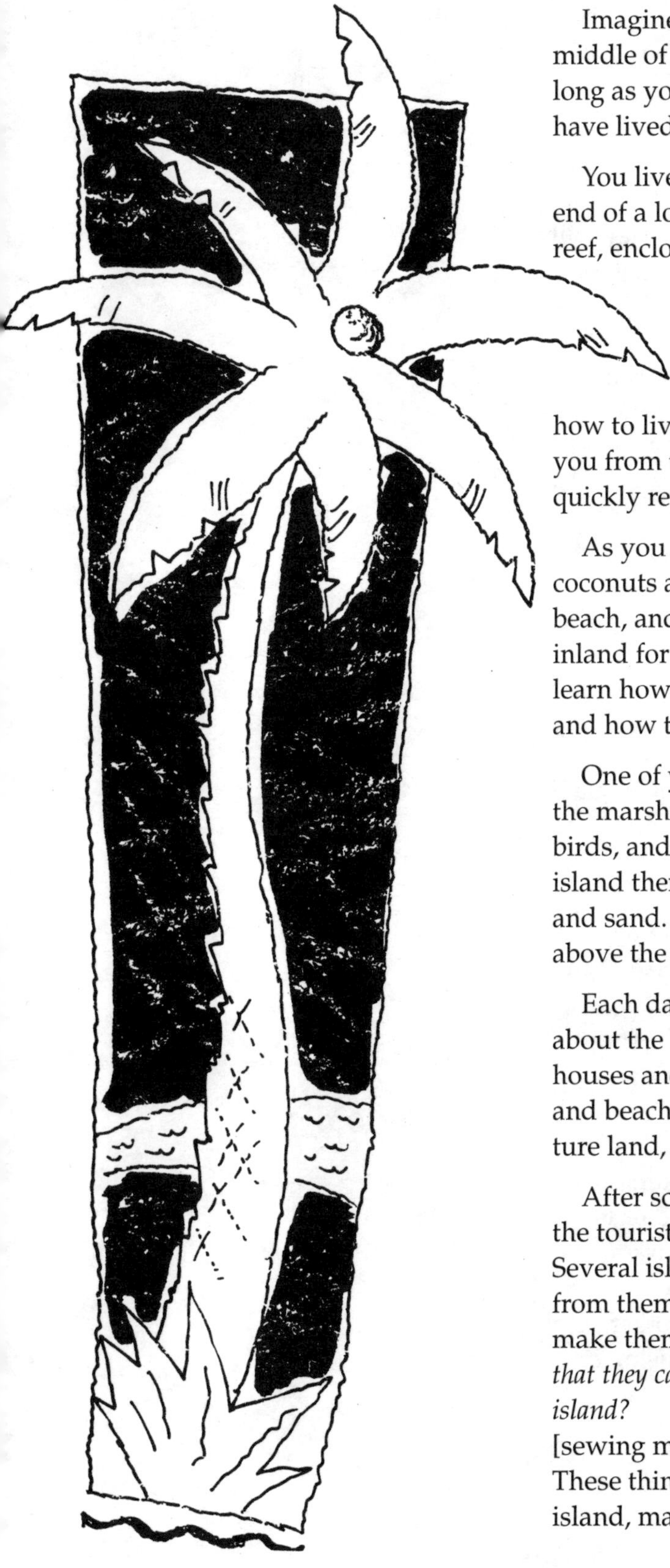

Imagine you live on one of a cluster of tropical islands, in the middle of the Pacific Ocean. Your name is Noua, and for as long as your elders can remember, the spirits of your people have lived on these islands.

You live in a village of wooden, thatch-roofed houses, at one end of a long white beach. Beyond the beach is a beautiful coral reef, enclosing a small lagoon of bright blue water. The weather is almost always hot and balmy—there is no real winter. But at a certain time in the year the hurricanes come. The hurricanes are frightening, but over many generations your people have learned how to live with them—how to construct safe shelters to protect you from the high winds, giant waves and floods, and how to quickly repair the damage.

As you grow up on your island, you learn how to collect coconuts and bananas from the trees along the back of the beach, and how to gather special plants from the forest further inland for eating and for making clothes and medicines. You learn how to grow vegetables on the fields next to the village, and how to hunt for fish in the lagoon and around the reef.

One of your daily tasks is to collect fresh water in tins from the marshy pool near the village. Here you find many water birds, and different sorts of fish, frogs, and tortoises. On the island there are no mountains to speak of, only tropical forest and sand. The highest point of the island is only a few feet above the level of the sea.

Each day you go to school in the village, but you also learn about the history of your island, and how to get food and make houses and clothes, from the old people. The lagoon and reef and beach and tropical forest are your playground, your adventure land, and your classroom.

After school, you earn some money by running errands for the tourists who stay in huts at the other end of the beach. Several island families own these tourist huts, and the cash from them is used by your village to buy the things they cannot make themselves. *What do you think people on your island need that they cannot make themselves from the natural resources on the island?*
[sewing machines, steel axes, medicines, refrigerators]
These things come from the stores in the big city, on a main island, many miles to the north.

You also learn many things from the tourists, and from the one TV in the village—about places where there are high mountains and huge buildings, and freeways with cars bumper to bumper, and bright lights, and hamburger stands, and air pollution. *How do you feel about these things?* It sounds very exciting, but also somehow scary and very big.

You are mostly happy with your life here, as you grow up among friends and sunshine and laughter, singing at night, and sharing grilled fish and coconut milk on the beach. Then, one night, there is a terrible hurricane, worse than in all living memory. Huge waves crash over the reef and sand bank, and sweep away the lower part of the village. Three people are killed, and many more are injured. Many families are left homeless, and have to be taken in by neighbors. The season's crops are destroyed, and some of the biggest coconut and banana trees along the beach are knocked down. Seawater has even swept into the fresh water pool, making the drinking water too salty to drink for weeks, so barrels of water have to be carried laboriously by canoe from a nearby island.

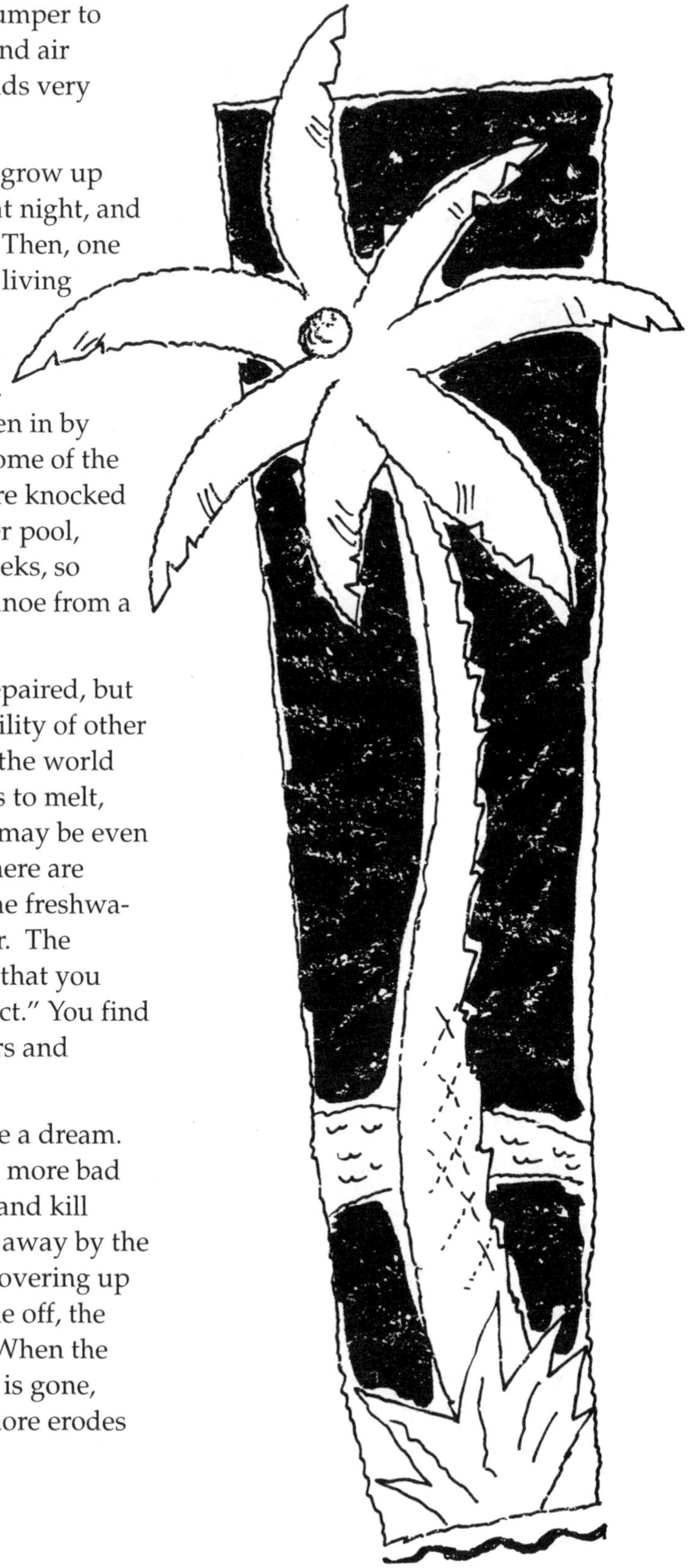

Over several months, the devastation is slowly repaired, but your people are now very worried about the possibility of other storms. There are reports on the TV about the way the world temperature is increasing, causing the polar ice caps to melt, and sea levels to rise. Some scientists predict there may be even more intense hurricanes in the coming years, and there are likely to be other effects on ocean islands, such as the freshwater pools being contaminated by the rising saltwater. The reports blame these changes on some strange thing that you haven't heard of before, called the "greenhouse effect." You find out that it is somehow caused by people driving cars and burning coal and oil, in strange countries far away.

One night after watching the TV reports you have a dream. In your dream the predictions come true—there are more bad hurricanes that destroy more of the village houses, and kill more people. The beautiful sandy beach is washed away by the huge waves and redeposited over part of the reef, covering up food for the fish. As the trees along the shoreline die off, the sand dunes at the top of the beach begin to erode. When the protection they gave to the plants behind the beach is gone, most of these plants turn brown and die, and the shore erodes even further.

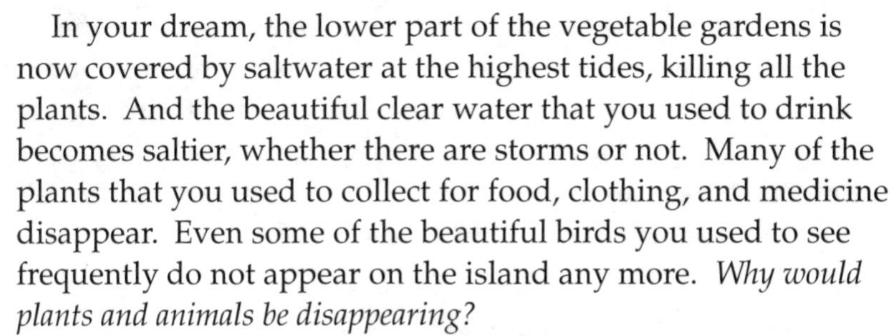

In your dream, the lower part of the vegetable gardens is now covered by saltwater at the highest tides, killing all the plants. And the beautiful clear water that you used to drink becomes saltier, whether there are storms or not. Many of the plants that you used to collect for food, clothing, and medicine disappear. Even some of the beautiful birds you used to see frequently do not appear on the island any more. *Why would plants and animals be disappearing?*

[Plants and animals are adapted to survive under a certain range of temperatures, but temperatures are beginning to change. Other changes in habitat are also occurring. For example, if there is less fresh water for humans, there is less fresh water for animals, too.]

Tourists have stopped coming to stay, and there is no money to buy spare parts for the sewing machine and refrigerator, or even to buy new axes. In the dream, a number of your friends have now grown older, and there is no work, or even food, to support them. One by one they are forced to leave the island, moving to the big city on the island to the north, along with young couples with children, who are no longer able to feed and clothe them on your island. When people leave, you hear very little about them, because communication with the big island isn't very good, and it seems that people easily get lost in the hugeness of the city. In the village people argue about why things are changing. Some of the old people say the spirits are angry with humans for something bad they have done.

Eventually, it is also time for you to leave. You will miss the celebrations when all the islanders got together to thank the spirits for a good season. You would have married and had children here, taught them how to catch fish on the reef, and plant in the vegetable patch next to the village. But now, that is not possible. So one sad day, you pack your few belongings, load them into a canoe on the beach you have always known, and kiss your parents and brothers and sisters good-bye. You paddle out past the lagoon and reef where you have spent your life playing, learning, and gathering your food, on your way to a nearby island, which has a plane to take you far away to the big city in the north.

At this point, Noua wakes up.

Pause for a few seconds so the students can reflect on the story. Ask,

If you were Noua, how would you be feeling?

What did you think about the story?

Name ____________________ Date ____________

AUTOMOBILE MANUFACTURERS

As representatives of automobile manufacturers around the world, you are concerned about profits for owners and investors, and jobs for workers. People who buy new cars seem to be mostly concerned about safety, engine power, and the cost of a new car. So, these are the guidelines you use to design them. In the past 10 years, many governments have required more effective pollution control equipment on cars and on the factories that produce them—all of which adds to the price of cars, making people think twice about buying a new one. Now, because of a predicted greenhouse effect, you are being asked to help reduce the amount of carbon dioxide by producing cars that are smaller, and car owners are being encouraged to use them only when necessary. You are worried about reducing profits. If this happens investors won't put their money into your companies, and you would have to close down some of your factories, which would put thousands of people out of work.

1. How would you describe the people you represent, and why they are concerned about the possibility of global warming?

2. List some questions or comments you might like to put to other groups at the conference.

3. List some ideas for what the people you represent can do to help cope with a warmer world.

4. List some ideas for what the people you represent can do to reduce the amount of carbon dioxide in the atmosphere.

Name ______________________________ Date ______________

ISLAND NATIONS

As representatives of the island nations, you are urgently concerned about rising sea levels, which are spoiling the freshwater and vegetable gardens on many islands. You are also very worried about the possible increase in devastating hurricanes, and the death of the coral reefs and mangrove swamps that your people rely on for fishing. You are very frustrated, because in order to solve your problem, people in large countries around the world must decide to produce less carbon dioxide and other gases. Most people in these other countries do not seem very interested in helping you, possibly because you represent less than 1% of the world's population. You are hopeful that people from large industrialized countries such as the United States will recognize that low coastal areas in their own countries also will be seriously affected by rising sea levels.

1. How would you describe the people you represent, and why they are concerned about the possibility of global warming?

2. List some questions or comments you might like to put to other groups at the conference.

3. List some ideas for what the people you represent can do to help cope with a warmer world.

4. List some ideas for what the people you represent can do to reduce the amount of carbon dioxide in the atmosphere.

Name ______________________________ Date ______________

AGRICULTURISTS

As representatives of the world agricultural community, including small farmers, large landowners, and livestock companies, you are caught in a dilemma. Major changes in the Earth's climate due to the greenhouse effect and global warming will affect which crops and animals are able to be raised in different regions. Thus, some farmers will be helped, and others may be devastated. You also know that animals such as cows, sheep, and horses contribute significantly to the greenhouse effect, through their production of methane. (Like carbon dioxide, methane is a greenhouse gas and it is estimated that it may account for about 25% of the global warming that is predicted to occur in the next century.) In addition, cutting down forests to make room for crop and pasture land also increases the amount of carbon dioxide in the atmosphere. While you worry about the consequences of these activities, you know that you must increase food production to feed the world's growing population.

1. How would you describe the people you represent, and why they are concerned about the possibility of global warming?

2. List some questions or comments you might like to put to other groups at the conference.

3. List some ideas for what the people you represent can do to help cope with a warmer world.

4. List some ideas for what the people you represent can do to reduce the amount of carbon dioxide and methane in the atmosphere.

Name ______________________________ Date ______________

CONSERVATIONISTS

As representatives of conservation groups around the world, you are very concerned about the major changes to the habitats of the world and to the unique plants and animals they contain, if the Earth warms up. You are also very concerned about the destruction of forests, especially rain forests in tropical and temperate areas that contain so many species. As many as 300–400 species of plants and animals become extinct each year due to the removal of forests. This "extinction crisis" is likely to increase with global warming. You would like to find ways of convincing people that the things each individual person does, such as driving cars, using plastic and paper packaging, even using too much electricity (which is made by burning fossil fuels) can have a detrimental effect on our global environment. You want people to realize that it makes good economic sense to conserve natural resources, whether or not the predictions about global warming turn out to be accurate.

1. How would you describe the people you represent, and why they are concerned about the possibility of global warming?

2. List some questions or comments you might like to put to other groups at the conference.

3. List some ideas for what the people you represent can do to help cope with a warmer world.

4. List some ideas for what the people you represent can do to reduce the amount of carbon dioxide in the atmosphere.

Name ______________________________ Date______________

WOOD & PAPER PRODUCERS

As representatives of the timber companies, wood mills, paper producers, and people who work in these industries around the world, you are proud that you provide people with useful products, such as wood to build houses and make furniture, and paper to write on, to publish books and newspapers, and to make containers. Unlike plastic products, containers made from wood are better for the environment because they decay and turn into useful soil. Yet some people blame you for destroying forests and contributing to the greenhouse effect. You already plant new trees when you cut down forests. You promote paper recycling programs, but not many people use these. Also, people seem to prefer to buy their goods in cartons rather than glass bottles that can be recycled. So why should it be your responsibility? You are aware that if the use of wood and paper products is reduced, thousands of people will lose their jobs.

1. How would you describe the people you represent, and why they are concerned about the possibility of global warming?

2. List some questions or comments you might like to put to other groups at the conference.

3. List some ideas for what the people you represent can do to help cope with a warmer world.

4. List some ideas for what the people you represent can do to reduce the amount of carbon dioxide in the atmosphere.

Following is an article about global climate change. Underline predictions about the **effects** that global climate change may have on the Earth's environment.

(Excerpted with permission from the San Francisco Chronicle, April 17, 1995.)

Name ______________________________ Date ______________

New Hints Of Global Warming

World temperatures are climbing again

by Charles Petit
Chronicle Science Writer

WHAT WE MIGHT EXPECT ON A WARMER EARTH

Here are possible consequences if the Earth's average temperature increases by several degrees in the next century, as many computer climate models predict.

SEA LEVEL

Warming expands ocean water and may melt some glaciers. Could go up one foot in next 35 years and two in next 100.

EXTREME STORMS

Hurricanes and other storms may become worse, but no clear sign of that yet.

DROUGHTS AND HEAT WAVES

Centers of large continents, such as the U.S. Great Plains, may be drier even if overall world rainfall increases somewhat. Heat waves may be more common.

REFUGEES

Movement of just 1 percent of a future population of 6 billion people due to higher sea level, drought, or other climate changes would produce 60 million migrants, many times the number of all refugees today .

AGRICULTURE

Impact mixed. Carbon dioxide stimulates plant growth. However, heat increases demand for water. Growing zones will shift if weather patterns change.

DISEASES

Warming that expands the tropics will also expand the range of tropical diseases such as malaria and other insect-born maladies.

WILDLIFE

Possible mass extinctions may occur as conditions change faster than species can move or adapt. Urban agriculture development leaves few wilderness corridors for migration.

Seven years after a sweltering heat wave helped make the "greenhouse effect" front-page news, scientists see new evidence suggesting that global warming is no far-off theory, but a real phenomenon that is already reshaping climate worldwide.

At the same time, they voice growing pessimism that the nations of the world will move anytime soon to reduce dramatically the human activities that are altering the chemistry of the protective atmosphere enveloping planet Earth.

After a brief cooling under the shade of volcanic dust spewed by Mount Pinatubo in 1991, world temperatures are again climbing to near-record levels. And even if it were possible to immediately freeze emissions of carbon dioxide—the chief "greenhouse" gas - levels high in the atmosphere would continue rising well into the next century.

"Is there global warming? I'm not 99 percent sure, but I am 90 percent sure," said Stephen Schneider, a climatologist at Stanford's Institute for International Studies.

Nobody can predict just what the long-term consequences of global warming might be. There could be good news—for instance, improved farm production at high latitudes as growing seasons lengthen. And there could be disaster-flooded coasts, droughts, outbreaks of disease, and mass extinctions as plants and animals expire before they can adapt or move.

On average, temperatures around the world are now about 1.2 degrees Fahrenheit warmer than in the 1860's when reliable record keeping began, and the changing patterns look increasingly like what the computers predict.

The best-known of greenhouse Cassandras is James Hansen of the Goddard Institute for Space Studies in New York. During the heat waves of 1988, Hansen caught the nation's attention when he told Congress he was convinced that global warming was under way.

The year 1990 brought a new record for annual average global temperature — 59.84 degrees Fahrenheit, compared with a 1951–1980 average of 59—and the next year was the second-hottest on record.

In 1992, haze from Pinatubo spread through the stratosphere, dimming sunlight at the surface slightly, and cooling the planet a bit, just as Hansen's computer models predicted. And, as he predicted, the cooling kicked global warming out of the news.

But Hansen's models also predicted the long-term warming would quickly resume. And, indeed, 1993 was the 10th-hottest on record (all 10 have been since 1973). Last year was the fifth-hottest, and so far, 1995 has been warm, too.

If there is one leading authority on global warming, it is probably the Intergovernmental Panel on Climate Change, a team of hundreds of the world's top climate experts appointed by the United Nations and World Meteorological Organization to weigh the evidence and forge a scientific consensus.

In 1990, and again in 1992, the panel projected that a doubling of carbon dioxide and other "greenhouse gases" will probably raise the average world temperatures 3 to 8 degrees Fahrenheit by late next century—a far faster increase than any seen in the geologic record going back tens of millions of years. One result, the group said, would be a two-foot rise in sea levels.

Like most climate experts, Michael MacCracken, an atmospheric scientist at the federal government's $1.8 billion Global Change Research Program, cautions that the jury is still out.

Following is an article about global climate change. Underline predictions about the **effects** that global climate change may have on the Earth's environment.

Name ______________________________ Date ______________

Human Activity and Climate Change

The "greenhouse effect" is so named because certain gases in the atmosphere trap heat and keep the earth warm, much as the glass of a greenhouse keeps the air inside warm. This atmospheric blanket is essential to life; without it, the earth would be much colder and uninhabitable. But civilization is now adding to the concentrations of 'greenhouse gases' in the atmosphere, apparently causing the Earth's temperature to rise beyond its natural level. Several gases are responsible, the most important being carbon dioxide (CO_2), which is released by the burning of coal, oil, natural gas, and by the destruction of forests. Scientists have little doubt that the increase in these gases will cause the earth to get warmer in the future, but exactly how much and how fast the temperature will rise is uncertain, as are the precise consequences of the rise. One almost certain change is that the oceans will rise, because warmer water will expand and Arctic ice sheets and alpine glaciers will partially melt. Analyses suggest the sea level rise will be between one and three feet by the mid-21st century, enough to cause severe coastal erosion, destroy irreplaceable wetlands, and contaminate water supplies and drainage systems with sea water. Major changes in weather patterns are also expected. Overall, average precipitation around the world should rise—but not necessarily where and when it is most needed. In the interiors of continents, the weather may actually become drier in the summer, causing more frequent droughts. And, as the oceans warm, the severity and frequency of tropical storms and hurricanes are likely to increase. Changes like these could have a serious impact on agriculture. In the United States, decreased rainfall and hotter summers in the Midwest and West could be devastating. Forests and wildlife could be destroyed, since many species may not be able to adapt quickly enough to changing conditions.

Excerpted from a pamphlet issued by the Union of Concerned Scientists, "The Heat Is On: Global Warming, The Greenhouse Effect & Energy Solutions." 1990.

Session 7: Worldwide Effects of Climate Change

Overview

The Effects Wheel activity is designed to communicate the concept of a chain of environmental effects. In this session, your students will create Effects Wheels to help them consider a wide range of impacts that might result from global warming. At the end of the session, the interest groups finalize their presentations to the World Conference, incorporating your feedback on their summaries from the previous session, and ideas from their Effects Wheels.

The purposes of this session are to:

(1) introduce the concept of a chain of effects that can result when a seemingly small change is made to a complex environmental system;

(2) provide students with the opportunity to explore a range of primary and secondary effects that may result from an increase in the global temperature;

(3) provide feedback and direction to the students as they prepare for the "World Conference on Global Warming" they will conduct in the next session.

What You Need

For the class

- ❏ 1 sheet of butcher paper
- ❏ 1 marking pen

For each group of 3 or 4 students

- ❏ 4 copies of the "Effects Wheel" sheet (see page 114)
- ❏ 1 pad of Post-Its™ (3" x 1½")
- ❏ 2 marking pens, of different colors (preferably green and red)
- ❏ 1 handout: "Flash! Messages" (see pages 115–117)

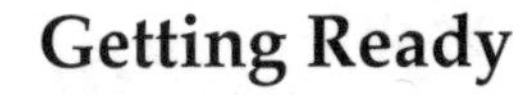

Getting Ready

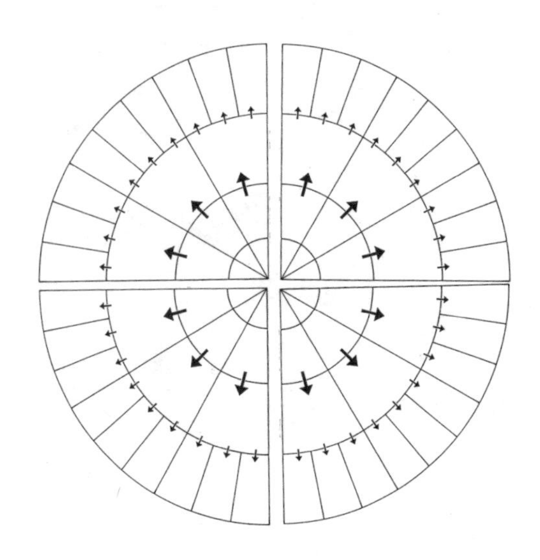

1. Photocopy the Effects Wheel sheet (see page 114). You will need four copies per group of three or four students. Cut the copies along the dotted lines so they are square, eliminating the extra paper outside the quarter circle on the long side of the paper (several sheets can be cut together, saving time). Use transparent tape to join the four sheets to make one large Effects Wheel. Set aside one per group of four students.

2. Read through the students' notes from Session 6. Make written comments and suggestions about how they might improve their presentations, and what other ideas and issues they might consider. These will be returned to the groups during this session to help them prepare for the World Conference.

3. Photocopy "Flash! Messages" sheets (see pages 115–117). Cut these apart, so you can give the messages to the appropriate groups near the end of this session.

The Effects Wheel Activity

1. Ask students to recall what predictions were made in the homework assignment article about the effects global warming might have. As students respond, make a vertical list on the board. If some effects named by the students follow from other effects, show the relationship with an arrow. For example:

   polar caps melt → sea level rises → coastal areas flooded

2. Point out that some of the effects listed occur as a result of other changes that happen before them—there is a "chain reaction" of effects. This makes it very difficult to predict all the effects of global warming.

3. Remind the class that most scientists investigating the greenhouse effect now agree that, if we do not change our ways of using energy, the average global temperature will increase, although there is disagreement about how much. Estimates range from 2°F to 7°F by the middle of the next century.

4. Inform the class that one way of thinking about the long term effects of a change, like a 5 °F rise in the world's average temperature, is to construct an **Effects Wheel**.

5. Explain that they are going to stay in the same groups, as in the previous session, to plan the World Conference. That way they can choose examples of effects of global warming that are important to their own interest group.

6. Hold up an example of the Effects Wheel, and explain that you want each group to write "+ 5°F" in the center circle, to represent a 5 °F rise in the average global temperature.

7. Explain that each group should choose four primary effects of global warming from those listed on the board (if necessary, explain that a primary effect is one that occurs as a direct result of the increase in global temperature), and write each of these in the segments adjacent to the inner circle.

8. Point to one of the four segments of the second circle and explain that this is where they should write a primary effect such as "sea level rises by three feet." Point to the segments of the circle just outside of this inner segment and explain that is where they would write the effects that would result from the sea level rise. Effects that would arise from those events would be written in the next outer circle, and so on.

9. Hold up a packet of Post-Its and explain that all of the effects are to be written on individual Post-It notes, and placed on the Effects Wheel. The packets can be pulled apart so **several students can write at the same time**. Explain that the Post-Its will allow the students to easily **change** their ideas if they want. Demonstrate how to write an effect on a Post-It and put it on the wheel.

10. Remind the students that they will hold a "World Conference on Global Warming" in the next session. In choosing which primary effects they put on their Effects Wheel, they may wish to consider those which will be of greatest interest to the people they represent.

11. Organize the class into conference preparation groups. Distribute one blank Effects Wheel, one pad of Post-Its, and pencils to each group. Ask if there are any questions, and then have the groups begin.

12. Circulate among the groups to assist as needed. After about 15 minutes, give each group a red pen and a green pen. Ask them to look at the effects listed on the outer circle of the wheel and to circle in green those outcomes they think are positive, in red those outcomes that are negative, and in black pencil those outcomes that are neutral. (Outcomes which are both positive and negative should be circled with both red and green.)

Sharing Completed Effects Wheels

1. After about 20 minutes (or sooner if all groups are finished), ask each group to report one of the primary effects they chose, and one positive and one negative outcome of this primary effect.

2. Ask each group to report how many positive, negative, and neutral outcomes resulted from their Effects Wheel. Make a table of these numbers as the groups report them. Ask the class to decide whether, on the basis of their predictions of the effects of a 5°F rise in global temperature, the outcomes would be mostly negative, or mostly positive, or a combination of both.

3. Ask each group to post their Effects Wheel on the wall of the classroom, and invite the students to look at all the Effects Wheels for ideas their group may not have considered. (Depending on the dynamics of your particular class and time available, you may wish to have groups send only one representative to inspect the Effects Wheels while the others are preparing for the conference.)

Preparing for the World Conference

1. Students should still be in their conference preparation groups. Return each group's notes from the previous session, along with your comments and suggestions. Tell them to go over their notes from the previous session, adding ideas that may have occurred to them since then. Tell them to also refer to their Effects Wheels.

2. Explain that at the World Conference each group should be ready to:

   a. **State their position** on dealing with global warming.

   b. **Ask questions or make statements** about other groups' positions.

   c. **Suggest solutions** that might be acceptable to most groups. (Summarize these tasks on the chalkboard).

3. Remind the students that the main purpose of the conference is **not to find someone to blame for global warming**, but rather to find out who and what is affected, and what can be done by **acting together**.

4. Hand out the "Flash! Messages" slips to the appropriate groups. These are designed to add some excitement to the conference preparation session, and to suggest some new ideas. Tell the students they do not need to use any of the ideas expressed in the messages if they do not want to do so.

5. During the small group discussions, circulate among the groups to offer suggestions. Some teachers have allowed groups to send representatives to lobby other groups as to the acceptability of some of their solutions.

6. Five minutes before the end of the session, remind students they should have clear answers to all of the questions, and should decide who in their group will speak first, second, and so on. Encourage the students to take turns speaking, rather than having just one student present the group's ideas. You might also want to plan with the class the order in which groups will present their ideas at the World Conference.

7. At the end of the session, collect each of the group's preparation notes.

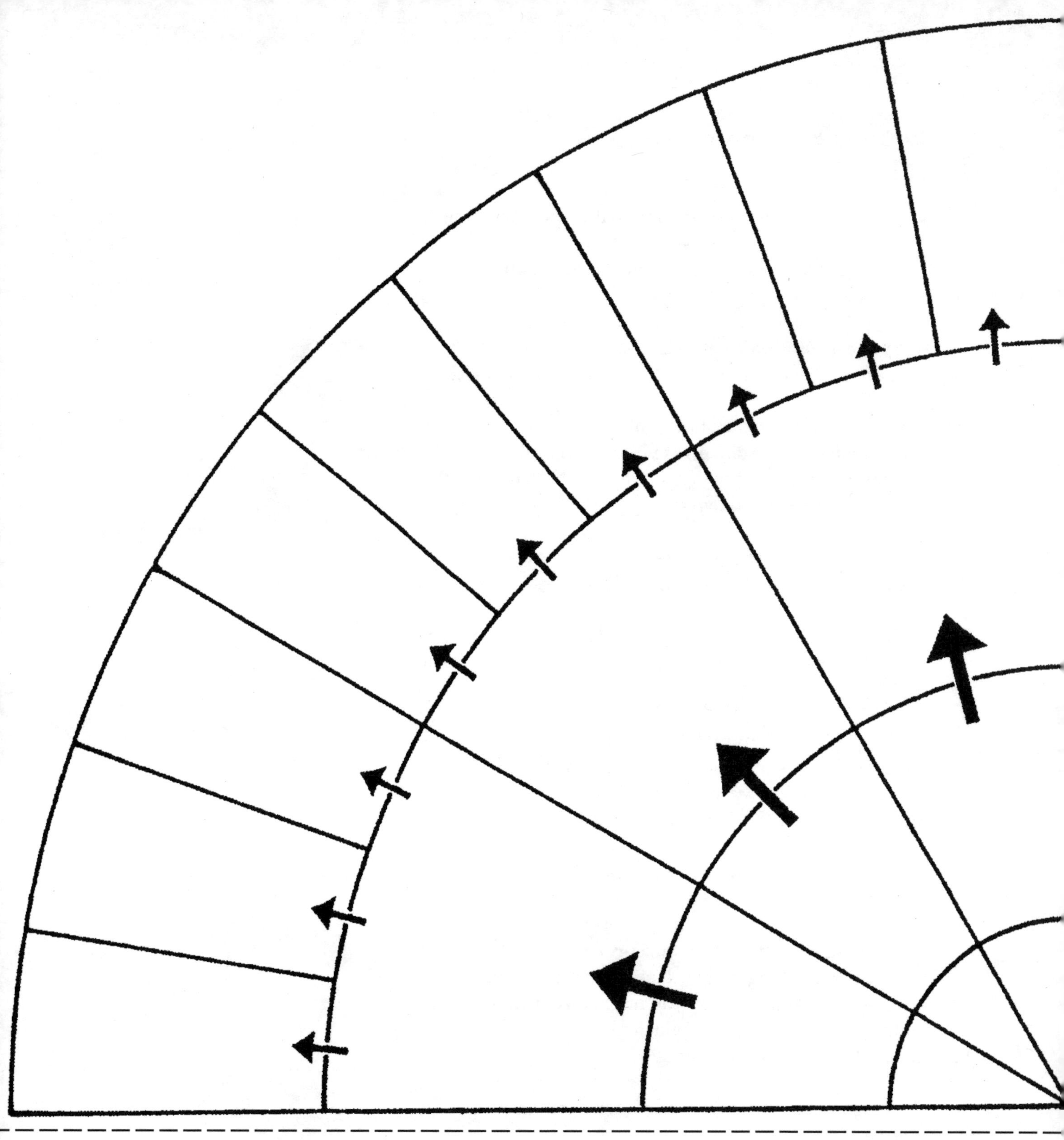

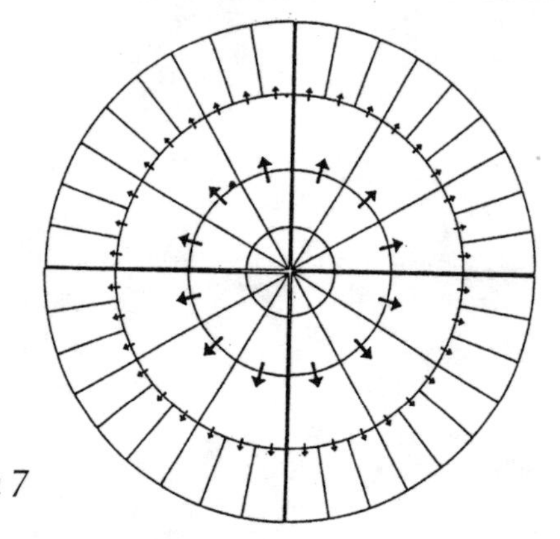

Cut along the dotted line and assemble as shown here.

FLASH!

MESSAGES FOR AUTOMOBILE MANUFACTURERS

- Your public relations office informs you the companies you represent have already reduced emissions of carbon dioxide by:
  1) improving the fuel efficiency of cars worldwide;
  2) installing energy-efficient lighting and heating in factories;
  3) encouraging employees to further conserve energy by turning off lights and machines when not in use.

  They want you to improve the image of auto manufacturers by telling people about these efforts.

- Your market researchers discovered the demand for large, heavy sport utility vehicles is growing rapidly. These vehicles have much poorer gas mileage than cars and are causing the amount of CO_2 concentration in the atmosphere of the United States to increase. You are urged to seek government funds for research into lighter, fuel-efficient vehicles that meet the needs of people who buy sport utility vehicles.

- Your Advanced Research Division is conducting research on alternative methods of transportation that greatly reduce the amount of carbon dioxide released in the atmosphere. One of these methods is Maglev trains, which are suspended above the tracks by magnetic fields. You would like more funds to greatly expand this line of research.

FLASH!

MESSAGES FOR WOOD & PAPER PRODUCERS

- Your market analysts found that the growing population of the world has increased the need for housing and paper products. This has increased the orders for timber and other wood-based products. Also, advances in literacy rates have increased the need and demand for books and newspapers. You may wish to ask for assistance from other groups at the conference to determine how some of these needs might be met without increasing the amount of logging.

- Your research division points out that durable homes built of wood actually reduce greenhouse gases because carbon is stored in the wood. Research should be focused on how to prevent fire and rot, both of which release the carbon stored in the wood.

- Your union representatives say that efforts to reduce logging will put thousands of people out of work. They suggest that some of these people might be put to work planting new forests, but they wonder where the money will come from, since these forests will not be harvested for at least 50 years. They want you to find a solution to this problem at the World Conference.

FLASH!

MESSAGES FOR AGRICULTURISTS

- Biologists have discovered that certain plants grow rapidly with increased carbon dioxide. Unfortunately, weeds grow rapidly too. You would like the industrialized nations to provide funds for continuing research to find useful plants that will outgrow the weeds.

- Energy engineers tell you that farming produces methane, which is a significant greenhouse gas. However, this problem can be turned into an advantage if the gas is collected and burned, and the energy produced is used to replace existing fossil fuel power plants. This will require funding from the richer nations of the world.

- Farm economists tell you that if world climate patterns change drastically, some farmers will lose all they have from droughts and storms, while others will benefit from increased rainfall and warmer weather. The economists suggest setting up a worldwide insurance program that would help balance the risk. Such a program will cost money, but you think it is important not to raise the cost of food because that will hurt poor people the most.

 -

FLASH!

MESSAGES FOR CONSERVATIONISTS

- Environmental scientists report hundreds of plant and animal species are becoming extinct every year as forests are cut down, and mining of fossil fuels destroys the land. Reducing these activities will not only reduce the threat of global warming, but will also save plants and animals from extinction.

- Biologists have discovered fast-growing trees that can be planted worldwide. Increasing the amount of forested lands will mean:
    1) more of the world's carbon dioxide would be stored in the wood and leaves of these trees;
    2) plants and animals would have more sheltered areas to enable them to survive.

- Your transportation consultants inform you that this conference is a good opportunity to urge countries to reduce greenhouse gases by developing more mass transit systems, and finding ways to locate people closer to their work so they do not have to travel as far.

FLASH!

MESSAGES FOR ISLAND NATIONS

- You are working hard to bring the problem of rising sea levels to the attention of other nations, especially those that produce most of the greenhouse gases. You would like more people to know about your problems, and more concern from other nations of the world so something will be done about them.

- You are requesting funds from the World Bank, to undertake projects that will reduce the effects of global warming. These projects include alternative fresh water supplies and food sources, and programs to resettle people on higher islands, where this is possible. Since the World Bank receives its money mostly from the industrialized nations, you need their support.

- You have thought of approaching individual companies that contribute most to the global warming problem, such as car and wood products manufacturers, to help prevent global warming now, while there is still a chance to save your islands. They could reduce their output of carbon dioxide, or use their profits to plant trees and preserve untouched forests that absorb carbon dioxide.

Troubling Waters

A Rising Tropical Sea Could Devastate Some Caribbean Island Nations, Says Geographer Orman Granger

Tropical island nations, especially in the Caribbean, could soon find their lifeline industries of agriculture, fishing, and tourism devastated by the greenhouse effect, says Orman Granger, professor of geography.

US states bordering the Caribbean, such as Louisiana and Florida, might feel the same effects but their economies would not suffer catastrophically, Granger said.

Speaking at the American Association for the Advancement of Science annual meeting last month in New Orleans, he recommended that island nations without delay adopt land-use laws to cut the probabilities that the ocean will swallow or batter new housing and industries.

Even more important is data gathering, he said. Without accurate contour maps, how can policy planners cope with rising sea levels? Rain patterns, current sea levels, and how buildings respond to surf and hurricanes all need more research too.

Greenhouse gases, especially carbon dioxide from fuel burning, are heating earth's atmosphere and are believed to be changing its climate.

Expected global temperature increases under current climate models would cause as much as a 20 percent increase in precipitation by the year 2030, leading to increased sea levels worldwide. And in the Caribbean, changes in storm patterns could mean more, stronger hurricanes and the damage accompanying them.

In 1979, hurricane David destroyed 80 percent of the houses in Dominica in the eastern Caribbean, and "in 1988, Hurricane Gilbert destroyed the Jamaican economy," Granger said.

Caribbean region economies are "bioproductivity-based," Granger said, depending on food crops, forestry, and fishing. Salt-water intrusion now threatens crops, Granger said, since with every sea-level rise, salt water also rises inland.

Air richer in carbon dioxide helps crops in most parts of the world, boosting productivity of corn, wheat, and barley, Granger said. But in the Caribbean tropics, it boosts mostly weeds, while extra warmth hatches more insect pests.

'Island nations without delay (should) adopt land-use laws to cut the probabilities that the ocean will swallow or batter new housing and industries. Even more important is data gathering. Without accurate contour maps, how can policy planners cope with rising sea levels?'

Tropical fruits, manioc, and bananas, —all Caribbean mainstays—do not benefit from more carbon dioxide, Granger said. Peasants, excluded by planters from the best lands, cannot choose their crops—they must grow "what can be bought and shipped to Europe," Granger said.

(Midwestern farmers in the US, on the other hand, have a wider choice for their markets, he noted.)

After World War II, many Caribbean countries, induced by external lenders such as the World Bank, opted to put money into tourism rather than industry.

That tourism relies, according to Granger, on "beautiful beaches" which are also narrow and will disappear as the sea rises. At the same time, marginal tourist spots on the US mainland will become warmer and more attractive, competing for the tourist dollar.

In tourist-dependent Barbados, Jamaica, Antigua, Martinique, and Guadeloupe, economies would be wrecked, but citizens could retreat inland for survival. But sea-level rise is a death knell for volcanic Dominica and Grenada because coastal settlements have nowhere to go but upslope, according to Granger.

Fishing, too—both for export and for sustenance—will be ruined by higher water. Storm-surge waves will scour river mouths, redistributing sediments and altering environments vitally important for fish and shellfish reproduction. Thus inland fisheries also would suffer losses.

Damage to other habitats such as mangrove swamps would threaten flamingoes and other birds, Granger said, as well as peasant fishing.

Granger could not speculate how open-sea fishing would be affected, but noted that ocean temperature changes will alter the mix of species.

A native of Trinidad, Granger has been a member of the faculty since 1974. His climatological research has focused on California, Greenland, the Caribbean, the American Southwest, and climatological theory.

—*Lynn Atwood*

Orman Granger (Jane Scherr photo)

Reprinted from the *Berkeleyan*, a newspaper for faculty and staff of the University of California at Berkeley.

Scientific studies, news reports, statements of urgent environmental concern, and international meetings about global warming and related issues have become highly frequent occurrences as the world enters the 21st century. Major events have included the Earth Summit in Rio de Janeiro, Brazil in 1992, the Kyoto Protocol in 1997, and more recent conferences in 2001 in Bonn, Germany, and Marrakesh, Morocco. All of these gatherings and others have been aimed at finalizing an international treaty on the reduction of greenhouse gases. You may want to encourage your students to investigate the current status of these efforts, and to consider the shifting policy of the United States, which withdrew from the Kyoto Protocol in March 2001.

Session 8: World Conference On Global Warming

Overview

This final session, in which your students conduct a mock "World Conference on Global Warming," provides an opportunity for them to synthesize what they have learned about the greenhouse effect and global warming scientifically, with what they think and feel personally, and how they should act as socially responsible citizens.

The discussion in this session needs to be tightly structured to run well, particularly if the class is not experienced in this type of activity. Although specific procedures and times for the World Conference are suggested, we encourage you to freely adapt them to suit your students. In any case, the discussion should have a definite conclusion with an opportunity for the students to express their personal opinions about how to respond to the global warming dilemma.

Even though the students are asked to focus on finding solutions, the purpose of the World Conference activity is not to find definitive answers—no one has been able to do that yet. Instead, the aims of this session are for the students to:

(1) recall what they have learned about the greenhouse effect and global warming;
(2) express what they have learned from a particular point of view;
(3) work with people who have different viewpoints to achieve solutions that all can agree upon;
(4) reflect on their own feelings about this issue;
(5) gain confidence in their abilities to participate in decisions that have global impact.

The book Global Warming: Are We Entering the Greenhouse Century? *by Stephen Schneider (see Resources, page 146) contains excellent and accessible scientific information, as well as revealing information and analysis of actual international conferences and other high-level meetings on the global warming issue. There is a discussion of the June 1988 conference in Toronto ("Changing Atmosphere: Implications for Global Security"), which was attended by more than 300 scientific, economic, legal, environmental, and governmental representatives from 46 countries. The role of the media in the global warming controversy, decision-making in a democracy, and practical suggestions for solutions originating from the general public are also discussed.*

Another book by Schneider, Laboratory Earth: The Planetary Gamble We Can't Afford to Lose, *is a 1997 update that takes into account recent achievements in climate modeling and the report of the Intergovernmental Panel on Climate Change, which makes detailed projections about climatic changes that are likely to occur by the end of the next century.*

What You Need

For the class

- ❑ 4 sheets of butcher paper
- ❑ 1 roll of masking tape

For each group of four students

- ❑ 5 sentence strips
- ❑ 1 or more marking pens

Getting Ready

1. Read over the notes on possible solutions students made in Session 7, and add your comments as needed.

2. Organize a space on one wall of the room where interest groups can post their solution slips.

3. Organize the seating in the room into a conference-style arrangement, such as a circle or semicircle.

4. You may want to do some extra preparation for the conference such as a banner to hang in the classroom announcing the "World Conference on Global Warming" and name cards for each of the groups.

GO!

Begin the Conference (10 minutes)

1. Invite the students to be seated in their interest groups for the conference. Hand out their notes from the previous session.

2. Introduce the session along the following lines:

   "As you are aware, in the last two sessions, a number of interest groups have been working to prepare their points of view to present to the World Conference on Global Warming."

   "Today, we have invited these groups to this international event, to hear their concerns about global warming, and to see if we can find some common solutions to this worldwide problem."

   "Let me go to the groups in turn, so you can tell us who you are."

3. After each group has introduced itself, announce: "Now we will give each group five minutes to finalize their statements to the conference, and solutions they may be able to suggest to either **cope with a warmer world**, or to **reduce the amount of carbon dioxide** in the atmosphere.

4. Instruct the groups to write each suggested solution on a separate sentence strip. Tell them to choose several people in their group to present their statements and suggestions.

5. Hand out five sentence strips to each group. As the groups are preparing, circulate to answer questions and offer encouragement. After five minutes, call the students back together to make their presentations to the conference.

Interest Groups Make Presentations (25 minutes)

1. Explain the procedure: "Each group will take no more than three minutes to make their statements and suggest solutions. Immediately following each group's presentation, I will allow two additional minutes for questions and comments from the other groups. I would ask that you state your points of view courteously, because one of the main purposes of this conference is to see how we can work together."

2. As possible solutions are suggested, they should be posted on the board or butcher paper.

3. At the conclusion of each presentation, ask if there are any questions or comments. Try to limit the time for each presentation, including discussion, to a total of five minutes.

4. After all groups have presented, thank each group. Point out that each group has a different approach to the issue of global warming (you may want to quickly summarize these for the class) but that you are confident the conference will be able to find some actions that all or some of the groups can agree to do.

Discuss Possible Solutions (10 minutes)

1. Ask the assembled delegates whether they think any of the solutions posted are acceptable to all of the interest groups, or at least to most of them. Remove these sentence strips from the list and place them in a central location for further discussion.

2. Lead a discussion about the relative merits of these solutions. You may want to assist the class to combine similar statements or modify the wording to arrive at a short list of three or four statements that are likely to be widely acceptable.

3. Taking each solution in turn, give the groups 30 seconds to confer about whether or not the people they represent would vote "Yes," "Maybe," or "No." Take one vote from each group, and record the results in a column next to the solution.

4. Announce the "winning solutions" as those that were agreed on by a substantial number of groups. Suggest that, in a real World Conference, debate might go on for weeks or months to change ideas and modify proposals so they would be acceptable to everyone.

5. Inform the class the conference is now at an end. Explain that, "What I want you to do now is to leave behind your point of view as a member of an interest group at a world conference and look at the solutions before us as an individual. What do you **personally** think about each of the main solutions suggested? As a class, we are going to take a second vote on each of the solutions, to see how acceptable they might be to us."

6. Tell the students to vote on each solution as follows:

   If you strongly agree .... Hold both hands up.
   If it may be a good idea .... Hold one hand up.
   If you're not sure .... Fold your arms.
   If it's a bad idea....................... Signal "thumbs down."

7. Summarize the class response to each solution in a few words.

8. Point out that a majority vote on issues of the environment may not necessarily be the best way, and is definitely not the only way, of making decisions. In some cases, when a small group of people in a community, or a small nation in the world, is much more adversely affected than others, steps other than a majority vote may be needed to prevent serious damage.

9. In some cases, standards already set by international or federal organizations or agencies may be violated, and the issue is taken directly to the courts, rather than the voters. In other cases, the issue is first called to our attention, not by a vote or an official, but by a few individuals who start studying and raising the issue in the mass media and books, or through protests and demonstrations.

10. Remind the class that it is difficult to predict environmental change. Regardless of what we might want to do, our environment has the ultimate power of veto—it may not want to do what we vote it should!

Going Further

1. A world conference on environmental issues, the "Earth Summit," was held in Rio de Janeiro, Brazil, in 1992. One of the leading stories from that conference concerned the role played by the United States government during prior negotiations on a "global warming" treaty and the United States refusal to sign a biodiversity agreement. There were numerous other issues that arose, both at the official conference and at an alternative gathering also held in Rio. Historically, the conference represents the growing world sense of urgency and seriousness surrounding the degradation of the environment. It included the largest number of world leaders assembled at any one time. You may want to have your students research and report on an aspect of this world conference, or debate the position taken, for example, on control of carbon dioxide emissions, by the European community, as compared to U.S. officials.

2. A day or two after the unit is finished, you may want to conduct a review by discussing the statements and questions the students listed on the first day in light of what they have learned about global warming since then. For questions that remain, groups of students could be assigned the task of investigating articles and reports, and compiling a response that is presented (either verbally or in writing) to the rest of the class.

   You may want to discuss with the class the importance of being able to study and analyze all sides of an issue, and to hear opposing viewpoints, before making decisions. Global warming is a worldwide concern, and it's necessary to consider alternative viewpoints if we are to do anything about it. At the same time, it is very important that citizens have a sense of their own power to make decisions—it is up to everyone to gain an understanding of the issues and realize their personal responsibility to take part in decision-making processes.

3. Have your students, as a class project, prepare and desk-top-publish a "Global Warming Fact Sheet" for distribution in the school and/or community, or publication in a school or community newspaper. This could include suggestions about how people can learn more and what people can do. Students could also write articles describing what the class did during the unit and summarizing conclusions.

4. Arrange to have a climatologist, a local spokesperson for an environmental group, or some other scientist or professional with specialized knowledge, make a presentation to the class exploring the many controversial issues involved in the topic of global warming. Students could work in groups to further define the issues of greatest interest and concern and to come up with challenging questions. Afterward, students could write a news article reporting on the presentation.

5. Encourage the class to keep reading about global warming—the solutions are still unfolding, and with the information they have gained in this unit, they may be able play an important role in deciding what should be done about it in the future.

Behind the Scenes

What Are the Potential Consequences of Global Warming?

According to the most recent (1995) report of the International Panel on Climate Change (IPCC), a group of 2,500 leading scientists from more than 100 countries, by the year 2100 the average global temperature of 59.5°F will increase from 2°F to 7°F. This doesn't sound like a very big increase at all! To understand what all the fuss is about, we need to examine the difference between climate and weather, and then consider the implications of a worldwide climate change.

Weather specifies the temperature, humidity, precipitation, and other conditions of the atmosphere at a particular location on a particular date. ***Climate*** refers to the average weather conditions for a region, or for an entire planet, over a period of about 30 years. One way to see the dramatic differences in habitability of different regions on the Earth that result from small changes in the climate is to look at past climate changes.

About 500 years ago, the average world temperature dropped by about 2°F, ushering in the "Little Ice Age." This was disas-

How do we know what the temperature of the earth was thousands of years ago?

In some parts of the polar ice caps, the ice is two miles deep. Cores taken from the ice pack show annual layering. Estimates of past atmospheric temperatures can be made by measuring the relative amounts of two slightly different forms, or isotopes of hydrogen in water, one of which is a little heavier than the other. When the climate is warmer, energy available for evaporation is greater, which causes a larger proportion of water containing the heavier isotope to appear in snowfall for that year.

trous for the Vikings, who had colonized Greenland when it was green. With only a 2°F drop in the global average temperature, Greenland became perpetually covered with ice and snow, and the Viking colonies collapsed. At the same time, many groups of Native Americans were forced by the change in climate to shift their way of life from raising corn and hunting deer, to transient hunting societies dependent on bison. George Washington's bitterly cold winter at Valley Forge dates from near the end of the Little Ice Age.

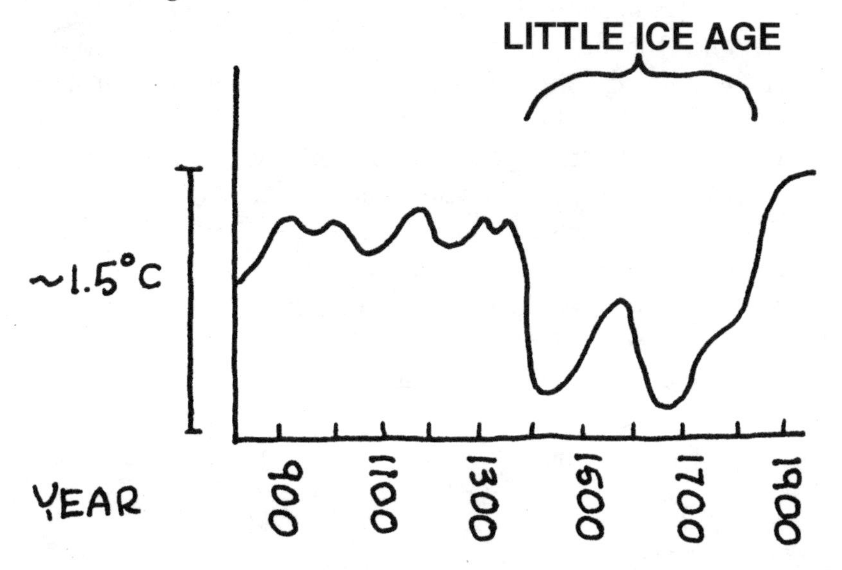

Graph adapted from *Wind, Water, Fire and Earth.* NSTA.

In the last major Ice Age (which ended about 10,000 years ago), the average global temperature was about 10°F to 14°F below the current average (59.5°F). During this period, the permanent polar ice sheets extended well below the U.S.-Canadian border, and the ice was more than a mile thick where New York City is today.

Worldwide climatic changes have taken place gradually in the past. The concern of climatologists is that as industrialization continues throughout the world, carbon dioxide and other greenhouse gases will continue to build up in the atmosphere at a rate of increase 10 to 50 times greater than when the Earth emerged from the most recent Ice Age. The IPCC report makes the following specific predictions of how this warming will affect Earth systems:

Ecosystems and Biodiversity

Ecosystems contain the Earth's entire reservoir of genetic and species diversity. They provide many goods and services critical to individuals and societies. The composition and geographic distribution of many ecosystems will shift as individual species respond to changes in climate. It is likely there will be reductions in biological diversity and the goods and services that ecosystems provide. Some ecological systems may not reach a new equilibrium for several centuries after the climate reaches a new balance.

Forests

The composition of forests is likely to change since the rate of climate change is expected to be faster than trees grow, reproduce, and reestablish themselves. With warmer temperatures, there are likely to be drier conditions, more extensive forest fires, and more frequent outbreaks of diseases and insect pests.

Rangelands

Increased levels of carbon dioxide may reduce the food value of some rangeland plants. Also, changes in seasons, precipitation, and average temperature are likely to change the boundaries between grasslands, forests, and shrub lands.

Deserts and Desertification

Deserts are likely to become hotter but not significantly wetter. Deserts are likely to expand, and new regions that are currently marginal may become deserts. Desertification is likely to become irreversible if the environment becomes drier and the soil is further degraded by erosion and compaction.

Ice Caps and Glaciers

Between one-third and one-half of existing mountain glacier mass could disappear over the next 100 years. The reduced extent of the glaciers could affect the water supply for hydroelectric power plants and agriculture. Melting of permafrost in certain areas could release large quantities of methane into the atmosphere, further increasing the concentration of greenhouse gases in the atmosphere. Little change is expected in the extent of the Greenland and Antarctic ice sheets over the next 50-100 years.

Mountain Regions

The predicted decrease in glacial ice and snow will affect rivers and soil stability. Vegetation is likely to shift to higher elevations and some species with climatic ranges limited to mountain tops may become extinct. Mountain resources such as food and fuel may be disrupted in many developing countries. Recreational industries—of increasing economic importance to many regions—are likely to be disrupted.

Lakes, Streams, and Wetlands

The effects will vary greatly by region. In some lakes and streams biological productivity would increase; while in other areas variability in water flow would result in more floods, droughts, and lowered water levels.

The Oceans

Sea levels could rise worldwide between 6" and 37", due to the expansion of water as it warms and the melting of glacial ice. Global ocean circulation patterns may change, which would affect climates in various parts of the world, as well as ocean productivity and fisheries. People who live in coastal areas or on islands where water quality is already a problem may find water supplies to be further reduced as saltwater from rising sea levels seeps into the fresh ground water, or due to increased droughts or floods.

Coastal Systems

The increase in sea level combined with an increase in the frequency and intensity of storm surges would erode beaches and coastal habitats, increase the salinity of estuaries and freshwater aquifers, change the tidal ranges in rivers and bays, and increase coastal flooding. Coastal ecosystems particularly at risk include saltwater marshes, mangrove ecosystems, coastal wetlands, coral reefs and atolls, and river deltas. Changes in these ecosystems would have major impacts on freshwater supplies, tourism, fisheries, and biodiversity.

Water Resources

Overall precipitation is likely to increase due to increased evaporation and cloud cover. However, as the pattern of precipitation changes, some regions that have abundant rainfall may experience droughts. Droughts have detrimental effects on agriculture, on river transportation systems, and on hydroelectric power. In addition to lower rainfall, if there are more hot days, soil evaporation will increase, exacerbating the effects of droughts. With generally warmer weather, some places will have rain instead of snow, increasing the chance of winter floods and reducing the amount of stored water in the ground for the spring and summer. In other areas, there may be an increased snow pack, resulting in spring flooding when the snow melts.

Agriculture

As the pattern of rainfall and soil moisture changes, it will affect agricultural productivity. In some areas rainfall is likely to increase, while in other areas it will probably decrease. For example, according to some models, agricultural productivity may decrease in the United States and increase in Canada. There may be increased risk of hunger and famine in some locations, such as sub-Saharan Africa, parts of Asia, Latin America, and some Pacific Island nations. Plant diseases and insects will also increase in many areas. The capacity of various countries to adapt to these changes will depend to a large extent on the speed of climate change.

Fisheries

Positive effects, such as a longer growing season and faster growth rate at high latitudes may be offset by negative factors such as changes in reproductive patterns, migration routes, and ecosystems.

Human Infrastructure

Coastal populations may be more vulnerable to flooding and erosional land loss. Estimates put about 46 million people per year currently at risk of flooding due to storm surges. In the absence of adaptation measures, a 50 centimeter sea level rise would increase this number to about 92 million; a 1-meter sea level rise would raise it to 118 million. If anticipated population growth is taken into account, the estimates increase even more rapidly. Estimated land losses are substantial for countries such as the Netherlands (6%), Bangladesh (17.5%), and Majuro Atoll in the Marshall Islands (80%). People who are forced out of low-lying coastal areas will migrate inland or to other countries, increasing the burden elsewhere.

Human Health

Direct health effects include more intense and longer heat waves, which are stressful for many people, especially those with respiratory diseases and the elderly. Indirect effects include an increase in the length of the season and geographical extent of vector-borne infectious tropical diseases such as malaria, dengue, yellow fever, and some viral encephalitis. For example, temperature increases in the upper range predicted by the IPCC could lead to 50-80 million additional malaria cases a year.

Surprise Impacts

It is unlikely that forecasters will be able to anticipate all of the effects of global climate change. Some of the surprises could be positive, such as the discovery of crops that grow well in an atmosphere rich in carbon dioxide. Others are likely to be detrimental, as people and other organisms struggle to adapt to the changes in climate.

What Are the Natural Causes of Climate Change?

The Goldilocks Effect

First Goldilocks visited Mars, and it was an average of 10°F below zero; far too cold! Then she visited Venus and it was an average of 840°F above zero; way too hot! Finally, she came to Earth and it was a comfortable average temperature of about 59°F; which was just right.

Over billions of years, living organisms have adapted to a wide range of temperature conditions on the Earth. However, so far as we can tell, life has not gained a foothold on either the intensely hot world of Venus or the freezing surface of Mars.

VENUS
840°F

For many years, climatologists have wondered why there are such great temperature differences between the three planets. Initially it was thought that Mars was much colder than Earth simply because it was further from the Sun; and Venus was much hotter because it was closer to the Sun. However, it is now known that the major factor in determining the average global temperatures of Venus, Earth, and Mars is not their distances from the Sun, but rather the amounts of certain gases in their atmospheres.

EARTH
59°F

In its early stage of development, Mars had a thicker atmosphere. It was much warmer than it is today, and there is strong evidence that water flowed on its surface. Over the past three to four billion years, Mars lost most of its atmosphere because it is smaller than the Earth and has much less gravity. With little atmosphere to trap its heat, it got very cold and its oceans and rivers dried up.

MARS
–10°F

Venus, on the other hand, is about the same size as the Earth. However, its atmosphere is composed almost entirely of carbon dioxide. The heat-trapping property of this gas is the main reason why it is hotter than a pizza oven on the surface of Venus.

The set of factors that has enabled the Earth to maintain conditions that are "just right" for life, while Venus is "too hot," and Mars is "too cold," has been aptly described as the Goldilocks Effect.

Let's take a closer look at our home planet, and see why it is currently just right for human habitation.

The Earth's Temperature—A Balancing Act

The sun's energy is radiated in all directions in the form of photons, or packets of solar energy. One property of photons is that they act like waves. We perceive photons with relatively short wavelengths as visible light when they strike our eyes. Other photons that have relatively long wavelengths are perceived as heat when they strike our bodies, and we call these photons infrared.

Photons with shorter wavelengths than visible light include ultraviolet, X-rays, and gamma rays. Photons with longer waves than infrared include microwaves and radio waves. Photons with shorter wavelengths have higher energy than those with longer wavelengths.

Only a small fraction of the huge number of the photons emitted by the Sun every second encounter the Earth. Of the photons that do encounter the Earth, about 30% are immediately reflected back into space. Another 24% are absorbed in the atmosphere; and about 46% are absorbed by the oceans and continents, making them warmer.

Imagine a single photon of visible light nearing the end of its eight-minute journey from the Sun to the Earth. By a happy coincidence it manages to miss dust particles and clouds in the atmosphere, and finally collides with a single molecule in a rock. The rock molecule is already vibrating because it has some heat energy. When the photon hits the rock molecule, the energy of the photon causes the rock molecule to vibrate even more. If we were to touch the rock, we would feel the vibration of the molecules against our skin as warmth.

Warm rocks soon cool off when they are no longer bombarded by photons (that is, in the shade or at night). If we could see a vibrating rock molecule cool off, we would see that it loses its energy by emitting a long wave infrared photon. If we placed our hands a few inches from the rock, we could feel such infrared photons leaving the rock as heat radiation.

If the Earth had very little atmosphere, like Mars, the rocks, soil, and oceans would still be warmed by sunlight. However, rapid cooling would quickly counterbalance this heating effect. The average surface temperature of the Earth with no atmosphere would stabilize at a little under 0°F. This point, at which the energy being absorbed equals the amount being given off is called ***thermal equilibrium.*** The Earth's equilibrium temperature is around 59°F. Why?

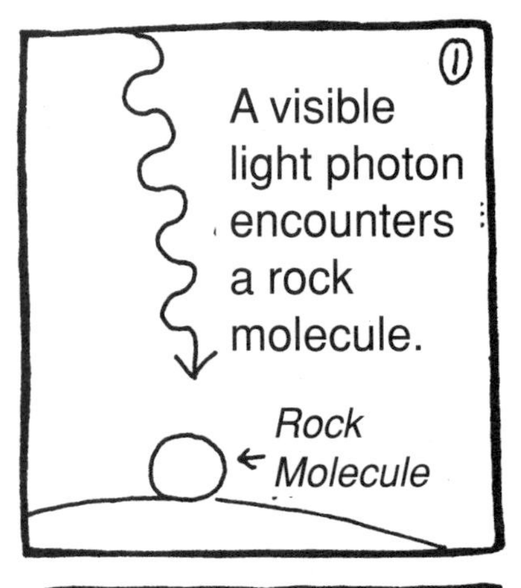

1
A visible light photon encounters a rock molecule.
Rock Molecule

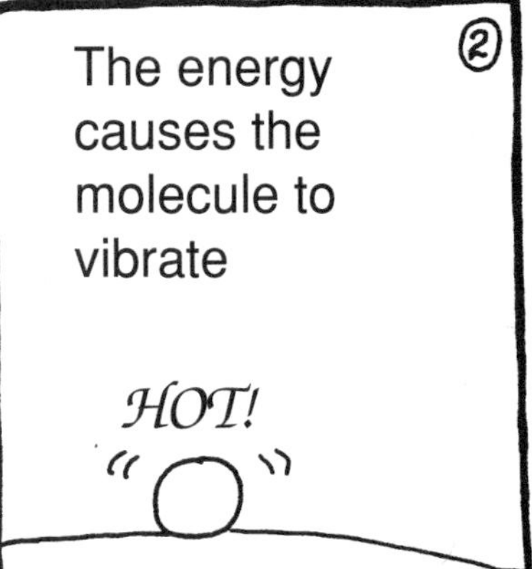

2
The energy causes the molecule to vibrate
HOT!

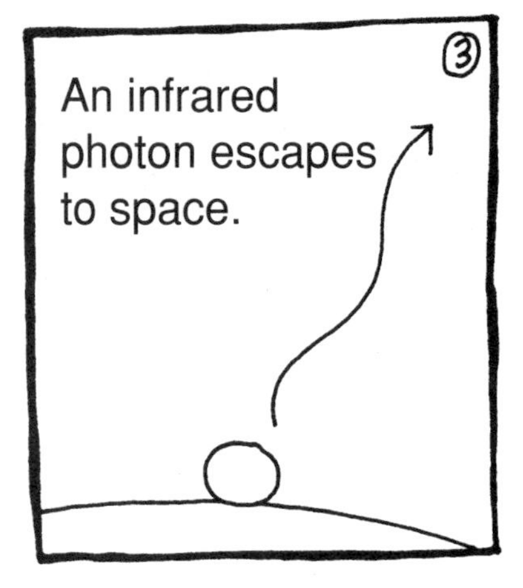

3
An infrared photon escapes to space.

A "Blanket" of Greenhouse Gases

The most common gases in our atmosphere, nitrogen and oxygen, are nearly transparent to both short wave and long wave photons. However, water vapor, carbon dioxide, methane, chlorofluorocarbons (CFCs), nitrous oxide, and ozone all absorb infrared photons easily. These gases are called "greenhouse gases" because they act like the glass in a greenhouse. That is, they are transparent to visible light, but they absorb infrared photons.

These gases make up only a few percent of the Earth's atmosphere, but account for just about all of its heat-trapping capacity.

Greenhouse gases trap heat by absorbing infrared photons the same way a rock molecule absorbs photons, by vibrating. They cool off the same way, by emitting an infrared photon. The infrared photon can go in any direction: upwards into space; sideways, where it may encounter another greenhouse gas molecule and warm the atmosphere further; or downwards, where it may again warm the Earth's surface.

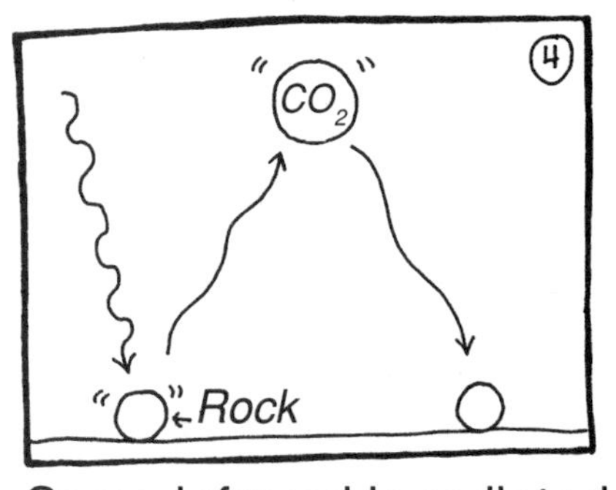

Some infrared is radiated back towards Earth.

Venus has a very dense atmosphere that consists almost entirely of carbon dioxide. The carbon dioxide molecules in Venus's atmosphere absorb and re-emit photons of infrared energy many times before the energy is finally allowed to escape into space. That is why the surface of Venus is so hot.

The Role of Earth's Oceans

Venus and Earth are called "twin" planets because they are very similar in size and composition. But what makes their atmospheres so different? One theory is that Earth formed an ocean very early in its history, while Venus did not. The water absorbed some of the carbon dioxide so that the planet cooled off.

That's probably part of the answer; but there's more. It became cool enough for life to evolve. According to fossil evidence, one of the first forms of life on Earth was blue-green algae. Blue-green algae absorb carbon dioxide and give off oxygen. Many paleoclimatologists (scientists who study past climates) believe that most of the oxygen in our atmosphere was produced by blue-green algae and other early microscopic life. Over thousands of millions of years, ocean waters and living things gradually absorbed much of the carbon dioxide in the atmosphere and replaced it with oxygen so that it became less and less like Venus.

When we take a more detailed look at our planet's climate history one other puzzle emerges. We know that there have been very warm and very cold periods in the distant past. When dinosaurs roamed the Earth much of the planet was covered with tropical forests, the oceans teemed with life, and there may have been no permanent polar caps as there are today. There have also been several ice ages, when the polar caps extended so far south that the New York City area was buried under a mile of ice. The puzzle is to explain how the Earth's climate was able to even out over the long run, so that it never became too hot nor too cold for life to survive. Over the past ten or fifteen years a theory has been developed to explain this. It's sometimes called the long-term carbon cycle.

Volcanoes and The Long Term Carbon Cycle

Volcanic eruptions are often accompanied by huge releases of carbon dioxide gas. It is possible that the warm periods of Earth's past were a result of periods when there were many volcanic eruptions. According to this theory, the volcanoes contributed massive amounts of carbon dioxide to the Earth's atmosphere and intensified the greenhouse effect. How, then, was the excess carbon dioxide eventually removed from the atmosphere so that the atmosphere cooled off?

Some of the excess was certainly absorbed by the oceans; but

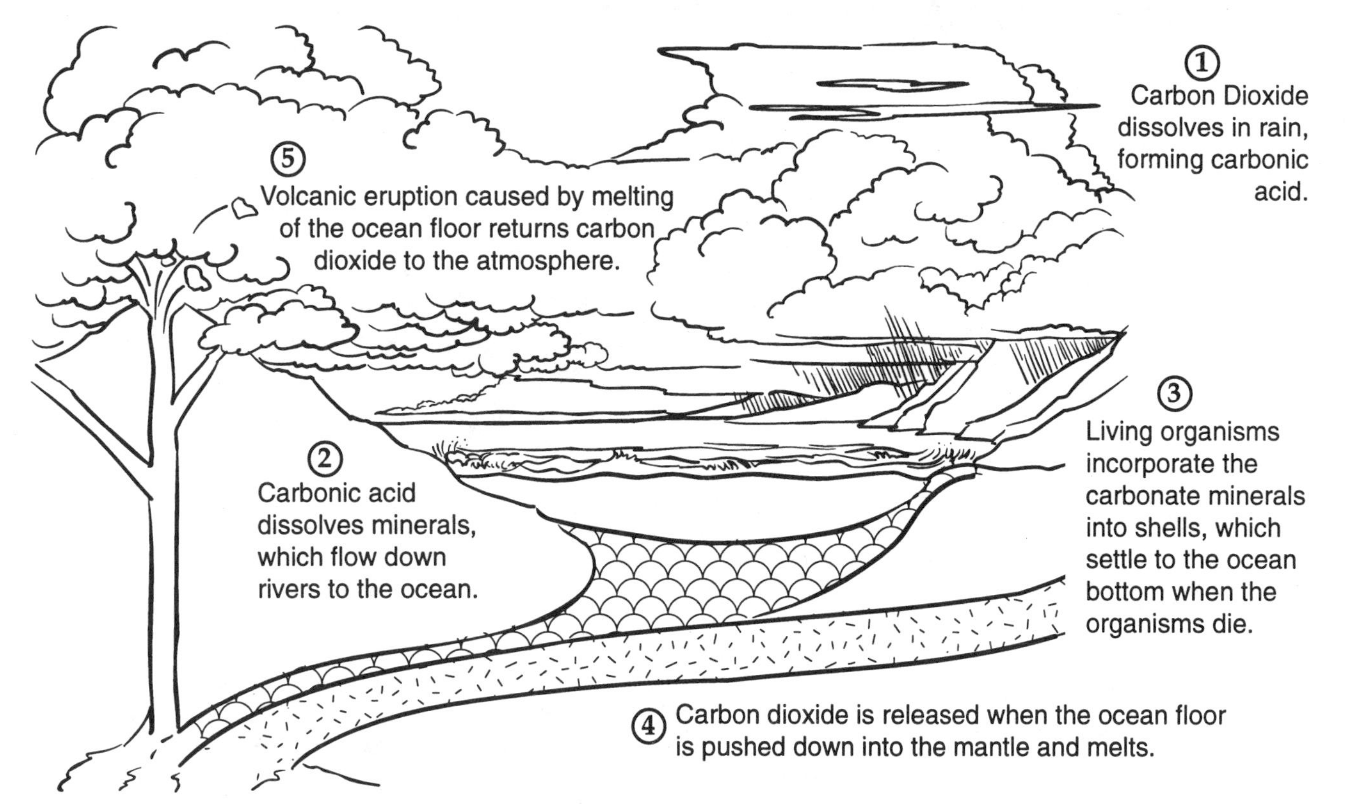
①
Carbon Dioxide dissolves in rain, forming carbonic acid.
②
Carbonic acid dissolves minerals, which flow down rivers to the ocean.
③
Living organisms incorporate the carbonate minerals into shells, which settle to the ocean bottom when the organisms die.
④
Carbon dioxide is released when the ocean floor is pushed down into the mantle and melts.
⑤
Volcanic eruption caused by melting of the ocean floor returns carbon dioxide to the atmosphere.

the world's oceans can absorb just so much. Another means of removing carbon dioxide was rain. Carbon dioxide dissolves into rain in the atmosphere, forming carbonic acid. This natural and very dilute acid rain falls and flows over and through the surface, where it dissolves minerals, forming compounds such as calcium carbonate. The mineral-laden waters flow into rivers, lakes, and oceans, and eventually form sedimentary rocks such as limestone.

The process of removing carbon dioxide from the atmosphere and trapping it in sedimentary rocks would have been greatly accelerated by the large numbers of marine organisms that thrived during very warm periods. These organisms would have incorporated dissolved carbon dioxide into shells and other hard body parts, then fallen to the bottom of the ocean where they build up in layers, forming sedimentary rocks.

If this process of trapping carbon into sedimentary rocks were to continue forever, the Earth would grow colder and colder. However, it is now widely accepted that the Earth's crust consists of a number of huge plates. Plates that make up the ocean floor tend to be pushed apart from spreading centers on the ocean bottom, and move towards the continents, where they are pushed down, or ***subducted*** under the continental crust. Pushed far under the edges of the continents, the rocks of the ocean floor eventually melt, and the carbon dioxide is again released, bubbling up to the surface where it is explosively released in a volcanic eruption! And the cycle begins again.

It takes tens of millions of years for a carbon atom to cycle through the atmosphere and oceans in the long term carbon cycle. However, not all carbon atoms are trapped in limestone. Some are absorbed by plants and cycled much more rapidly between the surface and atmosphere. This is the Short Term Carbon cycle.

How Are Humans Altering the Balance?

The rapid increase in the use of fossil fuels and other technologies over the past century is changing the balance of greenhouse gases in a very short time. The rate of change in the Earth's climate is estimated to be from 10 to 50 times as rapid as it was when the Earth emerged from the most recent Ice Age. This change in global temperature due to industrial, agricultural, and other human activity is commonly referred to as "global warming."

The concentration of the greenhouse gas methane, for example, has more than doubled as a result of widespread agriculture. As vegetation rots underwater in rice paddies, large quantities of methane are released. Grazing cattle also produce the gas in their digestive systems. The average cow releases up to 400 liters of methane per day. The increased termite activity in destroyed rain forests is believed to also add significant quantities of methane to the atmosphere.

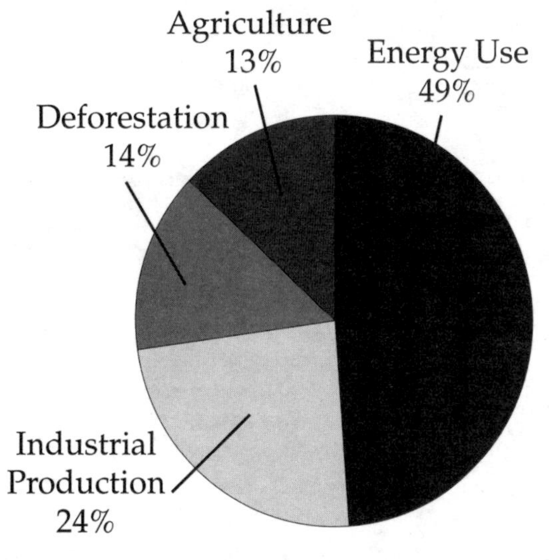

Source: World Resources Institute in collaboration with the United Nations Environmental Program, World Resources 1990–1991.

Methane is expected to account for about 18% of the increased greenhouse effect due to human activity. About 14% of the problem will come from chlorofluorocarbons (CFCs), which are products of the chemical industry; and another 6% will come from nitrous oxides, which come from auto exhausts. The pie chart shows which human activities contribute to global warming.

By far the largest contribution human activity makes to the Earth's greenhouse effect is through the copious production of carbon dioxide from power plants, automobiles, and virtually all industrial processes; and, both directly and indirectly, by the destruction of the rain forests and other deforestation. Carbon dioxide is expected to account for 49% of the increase in global temperatures due to human activity—as much as all of the other greenhouse gases together.

It is important to emphasize that a certain amount of carbon dioxide in the atmosphere is essential for our survival. Ever since the Earth was young, carbon dioxide has existed in the Earth's atmosphere. This carbon dioxide has kept temperatures in a range suitable for life.

The problem is the greatly increased amount of carbon dioxide in the atmosphere as a result of the industrial revolution that began in the 19th century, and has continued to expand throughout the world in the 20th century. In the early 1800s the concentration of carbon dioxide in the atmosphere is estimated to have been 275 parts per million (ppm). The current (1995) concentration is just under 360 ppm, an increase of about 25%.

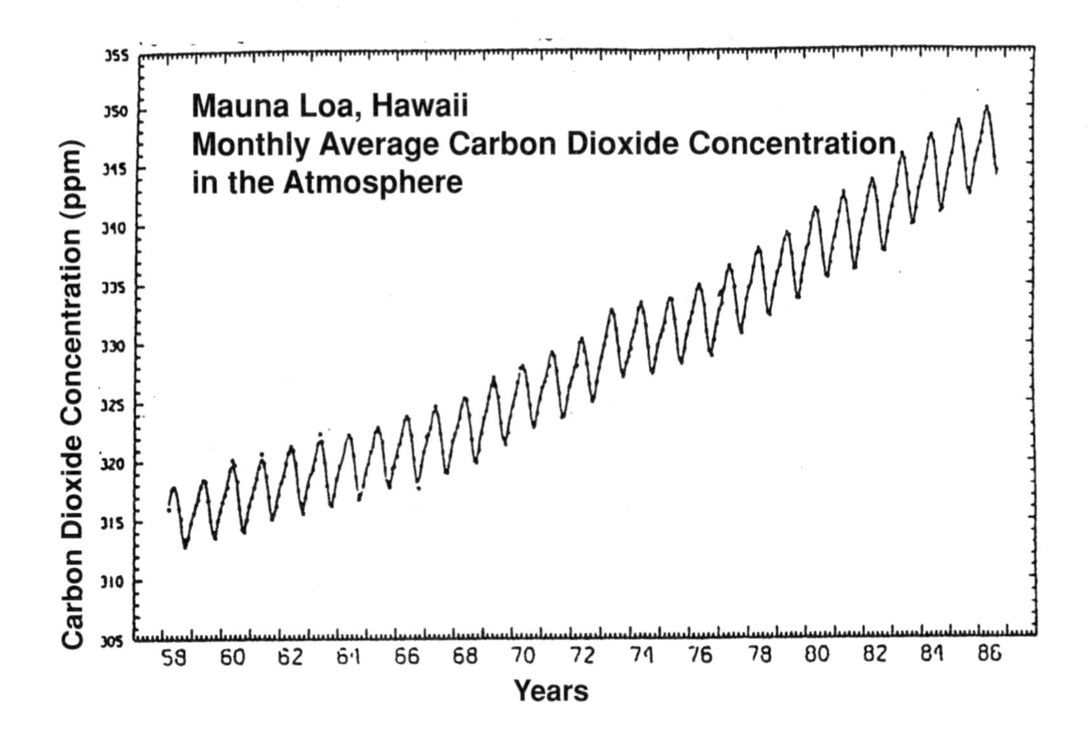

The Carbon Cycle in the Age of Industrialization

The diagram on page 137 shows how carbon is cycled around the biosphere of the Earth. The alphabetical letters below correlate to the boxes on the diagram.

A. Green plants take in carbon dioxide from the atmosphere. Using light energy from the sun, the plants rearrange the carbon atoms, to make long chained molecules with a "backbone" of carbon atoms. This process is called ***photosynthesis.***

B. Examples of molecules with carbon atom backbones, resulting from photosynthesis, include ***sugar***, which is transported around the plant to provide energy for growth; ***cellulose***, which makes up the structural support of most plants, such as wood; and ***proteins***, which are used in cells to control the complex chemical reactions necessary for life.

C. Animals and decomposers (such as bacteria and fungi) use the foods supplied by green plants as a source of energy. They release the energy plants harvest from the sun during photosynthesis by breaking the carbon atom connections and adding oxygen. This process is called ***cellular respiration***, and its main by-product is individual carbon atoms with oxygen attached—carbon dioxide. Animals and decomposers use the energy they get from respiration to make new connections between atoms (for growth and reproduction); for movement (within cells, between different parts of the organism, or to make the entire organism move); and to produce heat so important chemical processes within the organism can be kept at constant rates. Animals and decomposers also reorganize the carbon backbone molecules they get from plants to make the proteins, muscle fibers, and other tissues they need to grow, heal, and reproduce.

D. Occasionally in the past, when plants and animals died, their bodies were protected from decomposition, and the carbon backbone molecules that made up their cells were kept intact. Over time, these long chain carbon molecules underwent other chemical changes (depending on the type of organism and conditions under which they were stored) to form coal, oil, or other fossil fuels such as natural gas and peat. We extract these fossil fuels and burn them to get energy to run automobiles and factories, and to generate electricity. When we burn fossil fuels, we break the links in the long carbon chains that make them up, and release the energy stored there in a process similar to respiration. However, with burning, the energy is released much more quickly, and mainly in the form of heat (and some light). The separated carbon atoms join with oxygen in the air to form carbon dioxide, the same main by-product formed by respiration.

E. The current rate of increase of carbon dioxide in the Earth's atmosphere and oceans continues, both from natural processes such as respiration, decomposition, and forest fires, and from human activities, such as the burning of fossil fuels.

What Research Questions Are Still Unresolved?

How Much Will the Climate Change?

Detailed measurements and analyses of the Earth's atmosphere show beyond any doubt that greenhouse gases are becoming more abundant in the Earth's atmosphere. There is broad agreement that greater abundance of greenhouse gases means the average global temperature will increase.

Have we already seen evidence of global warming as a result of the increase in greenhouse gases?

The Earth's average temperature has increased from about 58°F in 1880, to about 59.5°F in the 1990s. 1990 and 1995 were the two warmest years on record. The major controversy surrounding this data is whether the observed warming of at least 1°F in the past 100 years is caused by the increase in greenhouse gases, or represents a natural variation in the Earth's climate, or a combination of both.

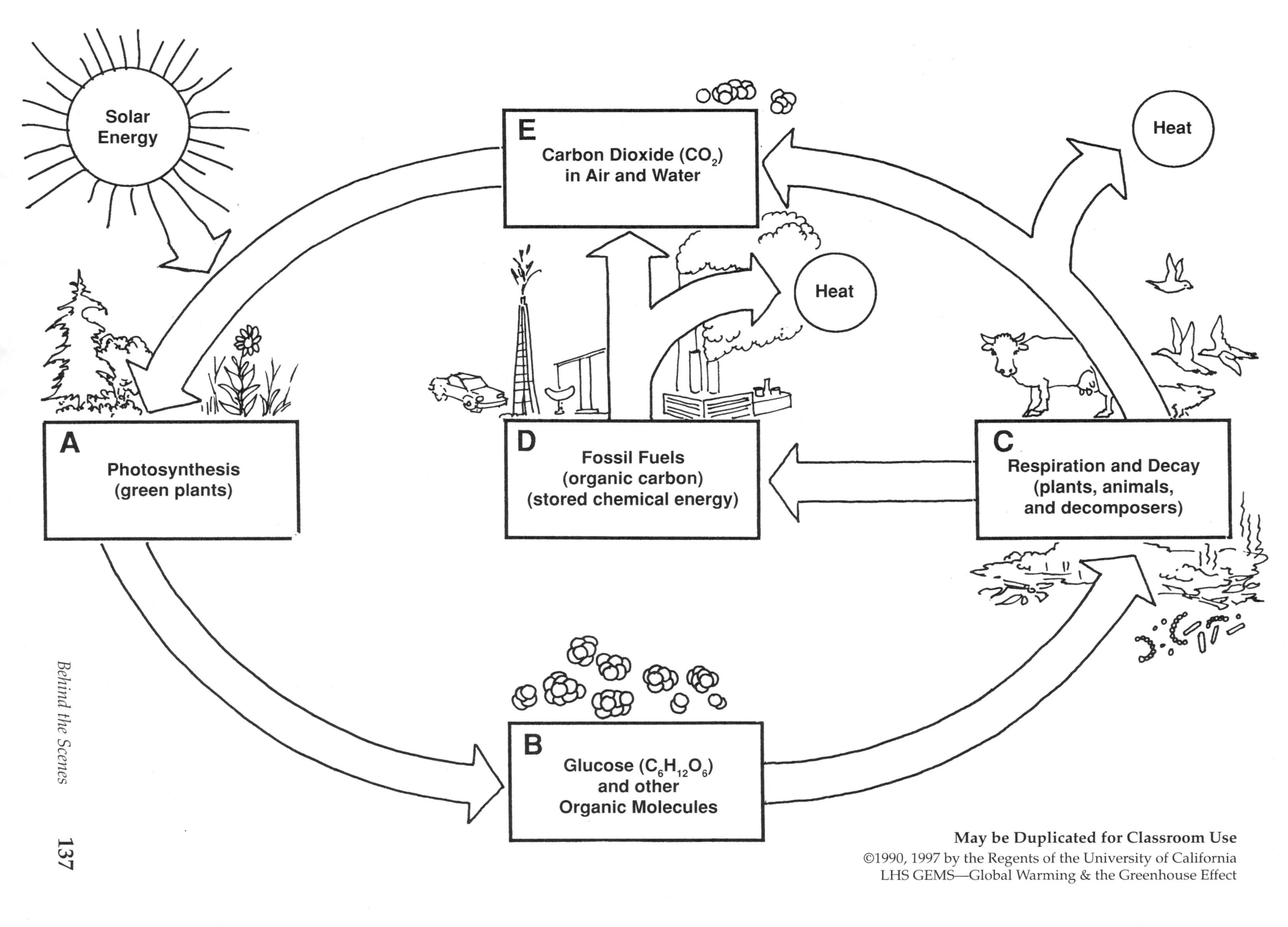

Solar Energy
E
Carbon Dioxide (CO_2) in Air and Water
Heat
Heat
A
Photosynthesis (green plants)
D
Fossil Fuels (organic carbon) (stored chemical energy)
C
Respiration and Decay (plants, animals, and decomposers)
B
Glucose ($C_6H_{12}O_6$) and other Organic Molecules

How rapidly will the concentration of greenhouse gases grow?

The answer to this question depends on many factors, but especially energy use in industry, transportation, and the home. For example, much of the carbon dioxide produced by automobiles could be reduced if people took advantage of mass transit or car pooling, worked at home, or lived near their work. Conservation of energy used in industries and homes could drastically reduce the number of fossil-fuel burning power plants that are built. In addition, reducing the destruction of the world's rain forests and reforesting other areas would significantly reduce the concentration of carbon dioxide in the air.

What natural processes may delay global warming?

The global systems that affect the concentrations of greenhouse gases in the atmosphere are not well understood. Some scientists believe that as the concentration of carbon dioxide in the atmosphere increases, the oceans and biosphere will become more effective in absorbing the gas. Others suspect that a warmer Earth will produce more clouds, which will reflect more of the sun's energy into space, cooling the Earth. These are examples of negative feedback effects.

What natural processes may speed global warming?

Some scientists expect increases in global temperatures to make changes in the Earth that will increase temperatures at an even more rapid rate. These are called positive feedback effects. For example, warming of the Earth will melt the polar ice caps, and smaller ice caps would be less effective in reflecting the sun's energy. Also, an increase in temperature is likely to stimulate bacteria and other decomposing agents to release carbon dioxide into the atmosphere at an accelerated rate.

What other factors cause climate change?

It is fairly well accepted that regular changes in the tilt of the Earth's axis and slight changes in the shape of its orbit around the sun resulted in the ice ages. These changes are very slow, and operate on scales of tens and hundreds of thousands of years. In contrast, the increased greenhouse effect due to burning of fossil fuels is occurring on the scale of decades.

What Can We Do?

It is unlikely we will be able to avert global warming entirely, given our current energy habits and dependence on fossil fuels. However, we can slow down the rate of change, perhaps giving us several additional decades to figure out how to adapt to changing world conditions. Here are some things we can do.

Conserve Energy

By far the most cost-effective solution to the potential problems of global warming is to use less energy. We have to burn less fossil fuel, so less carbon dioxide is added to the atmosphere.

Recent history shows we can make major savings through energy conservation. During the oil embargo of 1979-84, oil consumption per capita fell by one-sixth in industrialized countries (although energy consumption increased overall because of population increase). The economies of these countries remained relatively strong, despite predictions by some economists that economic growth depended on maintaining high growth in energy use.

Domestic energy efficiency can be increased. Purchase of the most energy-efficient appliances, lighting fixtures, and cars can significantly reduce the nation's energy use. Each of us makes many decisions daily involving energy usage. Every watt saved not only keeps money in our pockets, it also delays the development of global warming.

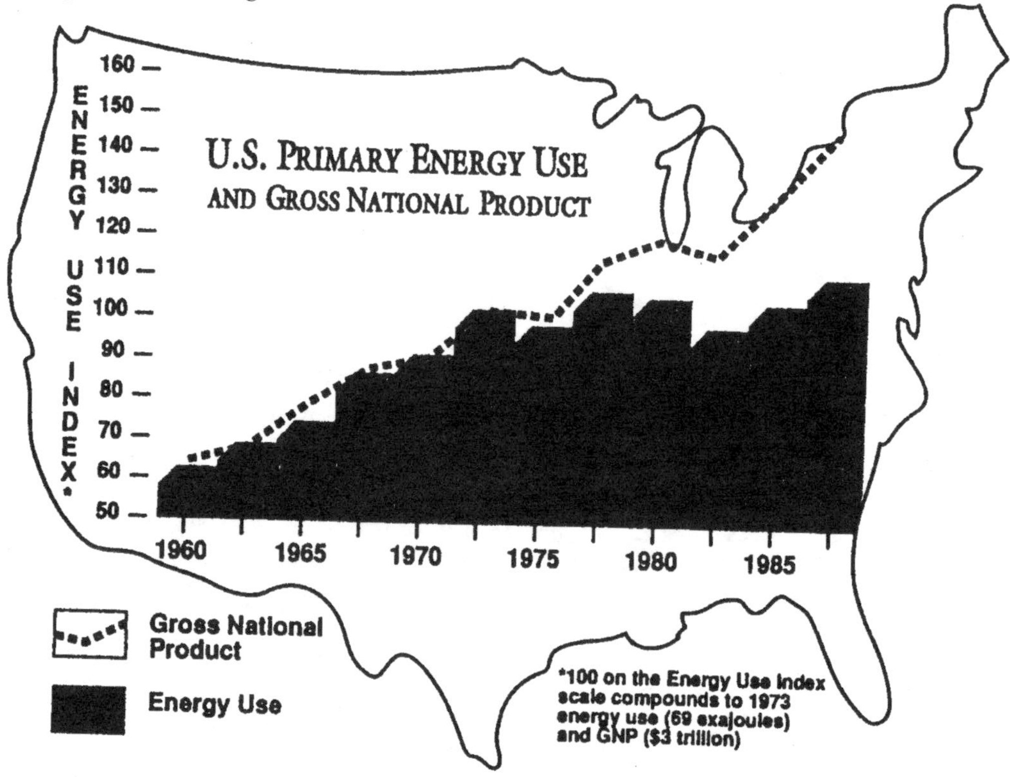

Recycle

A can of soda pop has a food energy value of 1 kilocalorie of energy. Yet the energy required to produce the aluminum can that packages it requires the expenditure of 2,200 kilocalories! Because most of the energy that goes into making a can is expended in refining the aluminum metal from crude ore, we can recover a considerable portion of that energy by tossing the can into a recycling bin instead of into the garbage. Or, we could avoid the dilemma altogether by using fully recyclable containers such as glass.

Communities across the nation are finding it increasingly difficult and expensive to dispose of their solid waste. Costs of incineration or burial are between $100 to $120 per ton. The cost for recycling a ton of material is about $40 per ton. Much of our solid waste consists of plastic or paper containers that produce carbon dioxide when they burn or decompose. We have the choice of recycling these materials into entirely new plastic and paper products, or replacing them with more durable containers that can be reused many times.

Discourage Deforestation

The most effective means of removing excess carbon dioxide from the atmosphere is by promoting forest growth. This is because trees use carbon dioxide to construct new wood. Rain forests lock up huge amounts of carbon dioxide. Unfortunately, the destruction of rain forests is reducing one of the world's major carbon dioxide reservoirs at the disheartening rate of over 50 acres per minute.

Leaders of certain developing nations, who are under fire for converting their rain forests into timber and agricultural land, claim that the industrialized world has already destroyed most of its own forests, and is now using wood from theirs. The historical record, which describes millions of square miles of forests that have been cleared in the United States, for example, supports their accusations. In addition, the United States is the world's second largest importer of tropical hardwoods, after Japan. From this perspective, the industrialized nations have a responsibility not only to preserve old growth forests within their own borders, but to contribute to solutions of the economic problems that have resulted in the destruction of rain forests in other countries.

Old growth (or climax) forests are the best reservoirs of carbon dioxide of any forest because they lock up the most carbon dioxide per unit of land area in their diversity of living organisms. The preservation of these forests will not only reduce the problem of global warming; it will also help to preserve thousands of species of plants and animals. These have intrinsic value in themselves. They provide a vast genetic pool of diversity for breeding new types of crops and animals, and offer new chemical structures for medicines and other uses. For example, about one-quarter of all pharmaceuticals have an active ingredient that originally derived from a tropical forest plant or animal.

Encourage Reforestation

Humankind has been eliminating forests for the last 10,000 years, but in some areas reforestation is underway. In some cases, as marginal agricultural land has been abandoned, forests have grown back. Local governments and nonprofit groups have undertaken tree-planting programs. In other cases, private timber companies have undertaken tree farming operations. Spectacular improvement in wood production has been achieved through careful species selection, genetic screening, spacing, thinning, pruning, fire, pest control, fertilization, and a harvesting cycle that encourages the growth of young, vigorous trees. These forest management practices help reduce atmospheric carbon dioxide, as long as they do not replace old growth forests. Additionally, timber used for building also adds to the planet's carbon storage capacity, as long as efforts are made to retard rot and prevent fire, and so long as the new buildings do not replace forests.

Alternative Energy Sources

As the Fossil Fuel graph on page 142 shows, the current period of fossil-fuel burning is a brief "blip" in the history of humankind. The rate of fossil fuel burning is increasing, not only in the industrialized nations, but in developing nations as well. For many reasons, in the decades ahead, we have to change our energy use habits. First, fossil fuels are a nonrenewable resource—when they are used up, they are gone forever. Second, toxic fumes produced when we burn these fuels (such as the oxides of nitrogen and sulfur) are polluting the atmosphere. And third, the released carbon dioxide is creating the potential for global warming.

Alternative energy sources do exist that will reduce the production of carbon dioxide, but at a price. The following alternatives require almost no burning of fuels, except to produce the required machinery, and thus produce very little carbon dioxide.

Wind Power

Electric generators driven by the wind are currently economical in areas where the wind blows almost continuously. Wind power reduces the production of carbon dioxide by 100% since no fossil fuels are burned.

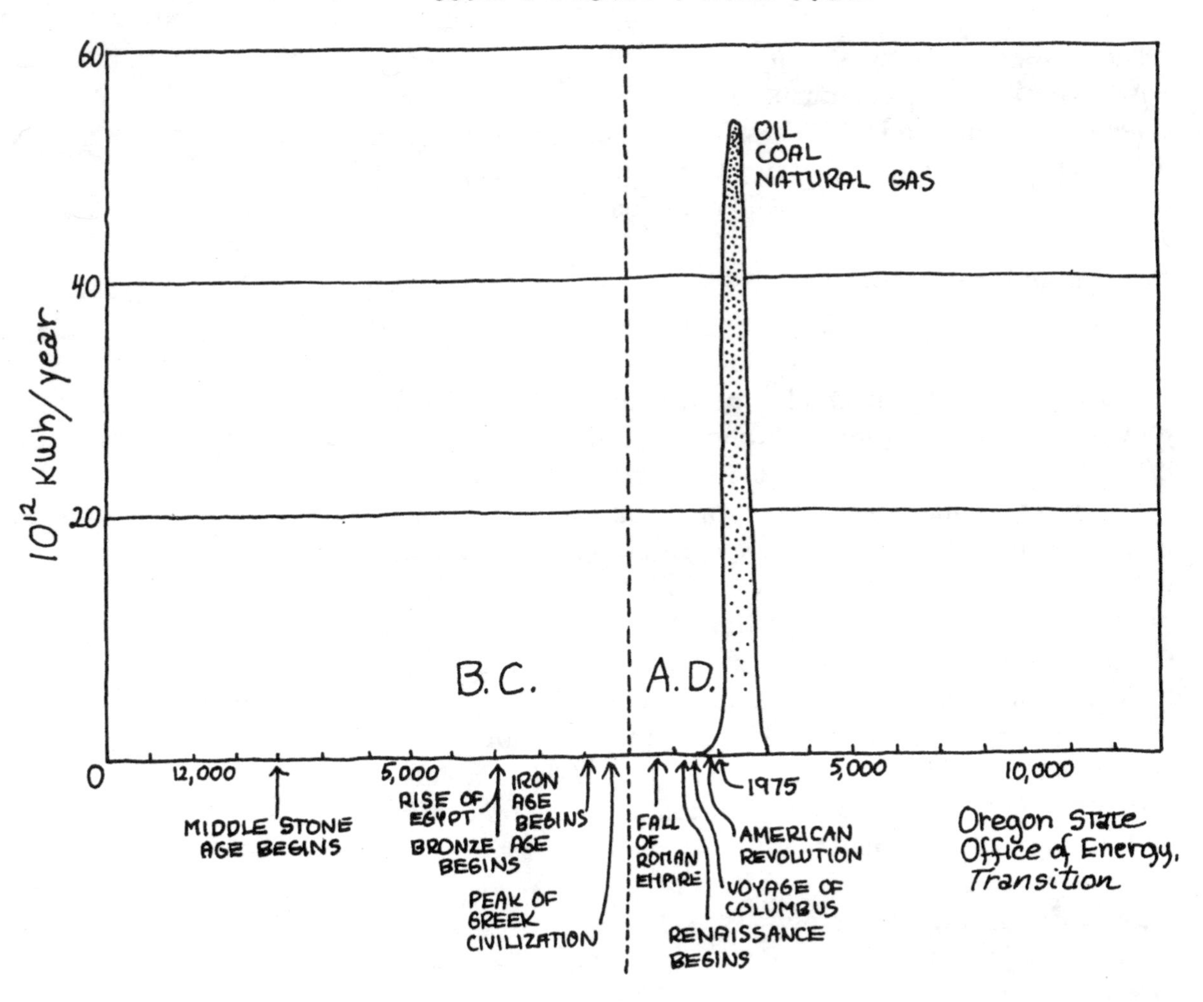

Geothermal Energy

Pressurized hot water from deep in the Earth can be tapped to produce energy in many regions of the world. Such systems do not add carbon dioxide to the atmosphere, but they do have their costs. Ironically, one controversial geothermal project is proposed for the island of Hawaii, where some of the remaining old growth rain forests will be destroyed to build the plant.

Solar Thermal Power

Large scale thermal power plants are currently being developed and tested in which mirrors focus sunlight onto tubes of oil. The hot oil is then used to run an electric generator.

Solar Cells

Also called photovoltaics, solar cells convert light energy directly into electrical energy. Solar cells are currently used in regions where there is lots of sun and it is too expensive to run lines from a distant large-scale power plant. The cost of electricity from solar cells is coming down, so it eventually may become a cost-effective way to produce electricity.

Passive Solar Heating

The greatest energy usage for individual households goes to produce hot water and control air temperature. Solar heaters can be built to trap the heat of the sun to provide hot water for a house or set of apartments. Windows, insulation, orientation, color of the house, and other design features can reduce the need for gas and electricity to heat the house in winter or cool it in summer. Although there was considerable interest in these approaches to reducing energy use in the 1960s and 1970s, tax incentives for installing these devices were removed in the 1980s. (The GEMS unit *Hot Water and Warm Homes from Sunlight* provides hands-on experiences with passive solar heating.)

Hydroelectric Power

Hydroelectric power is electricity formed by use of water power, generated by dams on rivers or even by using ocean tides. Hydroelectric power is a traditional and still important source of renewable energy that does not produce carbon dioxide. However, decisions to build new hydroelectric facilities must weigh the fact that energy is being produced without burning fossil fuels against the destruction of forests and other ecosystems. In many cases, the construction of a dam causes water to cover large areas of forest, agricultural land, or other heavily vegetated terrain that stores carbon dioxide.

Nuclear Fission

Nuclear power does not burn fossil fuels directly, and so does not release carbon dioxide into the atmosphere. However, the equivalent of about 10% of the energy produced by nuclear fission is required in the form of fossil fuels, used to mine the uranium ores needed for the fission process. The biggest problem with nuclear fission at present is that used nuclear fuel is radioactive. Finding safe storage for this toxic radioactive waste, where it has to be isolated from the environment for many thousands of years, is an unsolved problem.

Nuclear Fusion

In contrast to today's nuclear power plants that work by splitting uranium atoms (fission), nuclear fusion works by heating hydrogen to such high temperatures that two atoms combine (fuse) to form an atom of helium, giving off heat in the process. Although extensive research is being conducted, fusion power is still impractical. The chances for a cold fusion power plant at this juncture seem to be very slim.

The following alternatives require the burning of fuels, but would reduce the amount of carbon dioxide currently released in the atmosphere.

Garbage

Burning solid wastes to produce power will not contribute to the overall production of carbon dioxide, since the wastes would otherwise be burned, or would decay, thus releasing both carbon dioxide and methane. This approach solves waste-disposal problems and makes it unnecessary to burn additional fossil fuels. An even better alternative is to avoid producing excessive garbage by recycling.

Wood

Burning the output of tree farms also does not add to the overall carbon dioxide production, as long as the trees are replaced. Wood waste from the production of sawed timber is already burned to supplement the energy needs of the machines in some sawmills.

Methane

Power plants that burn methane produced by decomposing animal and vegetable matter can be very efficient at reducing global warming, even though carbon dioxide is the main product of burning. This is because a molecule of methane gas has about 25 times as much heat trapping capacity as a molecule of carbon dioxide.

None of the above alternative energy sources are yet able to replace fossil fuels, not only due to technical reasons, but also because of economics and our current energy habits. Yet we know that in the next few decades we have to wean ourselves from dependence on fossil fuel. We can delay that process by changing our behaviors to conserve energy; but eventually we will have to develop new forms of energy production, or make better use of the alternative energy sources that have already been developed.

Reduce Emissions of Other Greenhouse Gases

Chlorofluorocarbons

The chemicals used in refrigerators and as solvents in the computer microchip industry are not only contributing to the destruction of the ozone layer, they are also powerful greenhouse gases. Efforts to replace them with other chemical compounds are already underway, and production of CFCs is being phased out around the world.

Vehicle exhaust

The exhaust from cars and other vehicles contains not only large amounts of carbon dioxide, but oxides of nitrogen, also greenhouse gases. Efforts to reduce emissions through more efficient engines and use of alternative fuels (such as natural gas, gasohol, and hydrogen) will help slow global warming. Also important are efforts to reduce miles driven through carpools, mass transit systems, and working at home when possible. Unfortunately, the recent popularity of Sport Utility Vehicles has reversed the trend towards more gas-efficient cars. The United States is contributing more and more to global warming as these large, heavy vehicles become more and more popular.

Conclusions

Many of the important questions associated with global warming fall at the intersection of science and society. Since there is not full agreement in the scientific community about how much the globe will warm in the years ahead, the decisions that must be made are not easy.

Some people argue that we should take no action at all. After all, development of new power sources costs money. Deciding not to cut down old growth forests in Oregon, or calling a halt to oil drilling off the California coast, will reduce the number of jobs available. Why not just wait and see?

Other people say we cannot afford to wait. If we are to slow global warming enough to allow ourselves time to figure out how to adapt to the changes, we must start to do the following:

1) Conserve energy now.

2) Push ahead with research into new energy technologies and reforestation efforts.

3) Continue to monitor the environment and improve our abilities to predict and adapt to global changes.

The IPCC report calls these "no regrets" policies. That is, they will have major positive effects on the environment whether or not the worst predictions of global warming are realized. For example, protecting rain forests and other old growth forests will maintain the diversity of life. Conserving fossil fuels will improve the quality of the air, reduce oil spills and other damage to the environment, and keep available raw materials for making other petroleum products. And, changing our energy habits at home and where we work will allow us to save money.

The choices today's students make could have a profound impact on the quality of their lives and the lives of future generations. The issues involved are truly global and potentially affect all life on Earth.

Resources

Books

Energy and Climate Change:
Report of the Multi-Laboratory Climate Change Committee
Michael MacCracken, Chair
Lewis Publishers, Chelsea, Michigan, 161 pages, February, 1990.

Future Weather and the Greenhouse Effect
by John Gribbin,
Dell, New York. 1982.

Global Warming: Are We Entering the Greenhouse Century?
by Stephen H. Schneider
Random House (Sierra Club Books), New York, 1989.

Laboratory Earth: The Planetary Gamble We Can't Afford to Lose
by Stephen H. Schneider
Basic Books, HarperCollins, New York, 1997.

Booklets, Magazines, and Articles

"A Matter of Degrees," Irving Mintzer, World Resources Institute, Research Report #5, April 1987.

Atmospheric Carbon Dioxide and The Greenhouse Effect
U.S. Department of Energy, 1989. Available from the National Technical Information Service, U.S. Department of Commerce, Springfield, Virginia 22161.

"Endless Summer: Living With the Greenhouse Effect," Discover, October, 1988.

Global Warming, 1989
The Sierra Club, Public Affairs, 730 Polk St., San Francisco CA 94109.

"Global Warming Trends," Philip D. Jones and Tom M.L. Wigley, Scientific American, August 1990, pages 84-91.

"The Great Climate Debate," Robert M. White, Scientific American, July 1990, pages 36-43.

"The Greenhouse Effect," Science Activities, Volume 23, Number 1, February/March 1986.

The Greenhouse Effect: Implications for Economic Development
Erik Arrhenius and Thomas W. Waltz, World Bank Discussion Papers, The World Bank, Washington, D.C.
1990.

"The Heat Is On," Time, October 19, 1987.

The Heat Is On—Global Warming, The Greenhouse Effect, and Energy Solutions
The Union of Concerned Scientists, 26 Church St., Cambridge, MA 02238.

Lessons from the Rainforest: Essays by Norman Myers, Randall Hayes, Frances Moore Lappé, and others, edited by Suzanne Head and Robert Heinzman, Sierra Club Books, 1990.

"Managing Planet Earth," Scientific American, Volume 261, Number 3, September, 1989.

"Mission to Planet Earth," T.E Malone, Environment, Volume 28, Number 7, September, 1986.

"Our Changing Planet: The FY 1996 U.S. Global Change Research Program." A Report by the Subcommittee on Global Change Research, Committee on Environmental and Natural Resources Research of the National Technology Council, Supplement to the President's Fiscal Year 1996 Budget. Available from Global Change Research Information Office, 1747 Pennsylvania Avenue NW, Suite 200, Washington, D.C. 20006.

"Special Issue on Global Change," MOSAIC, Volume 19, Number 3/4, Fall/Winter, 1988, The National Science Foundation.

"Tropical Deforestation," Peter H. Raven, The Science Teacher, September, 1988.

Books for Young People on Global Warming

There are a large number of books published on global warming. They vary in tone, amount of scientific background, and estimate of possible consequences. You may want to emphasize to your students that their task, regardless of a particular author's bias, is to gather information to make their own independent judgments. Within the approximate age range for which this teacher's guide is intended, the following books may be of interest to your students.

The Greenhouse Effect: Life on a Warm Planet
by Rebecca L. Johnson
Lerner Publications Company, Minneapolis, Minnesota, 1990

Global Warming, Assessing the Greenhouse Threat
by Laurence Pringle
Arcade Publishing: Little, Brown and Co., New York, 1990

Our Global Greenhouse
by April Koral
Franklin Watts, New York, 1989

The Climate Crisis: Greenhouse Effect and Ozone Layer
by John Becklake
Franklin Watts, New York, 1989

The Greenhouse Effect
by Tony Hare
Gloucester Press ("Save Our Earth" series), New York, 1990

Kid Heroes of the Environment:
Simple Things Real Kids Are Doing to Save the Earth
edited by Catherine Dee; illustrated by Michele Montez
Earth Works Press, Berkeley, Calif., 1992
Photographs and concise descriptions of young environmental activists and their successful projects.

Posters, pamphlets, and newsletter

"The Greenhouse Gas-ette," newsletter available to teachers free of charge from Climate Protection Institute, 5833 Balmoral Dr., Oakland, CA 94619.

Global climate change posters and pamphlets are available from the National Science Foundation. Write to their National Science & Technology Week Office: NSTW '90, 1800 G Street, NW, Room 527, Washington, D. C. 20550.

Videos, Filmstrip, and Photos

"Greenhouse Crisis-The American Response," VHS 11 minutes, produced by the Union of Concerned Scientists, available from The Video Project, 5332 College Ave., Suite 101, Oakland, CA 94618.

"Our Threatened Heritage," VHS (19 minutes), National Wildlife Federation.

"Our Changing World," filmstrip available from Customer Service, Educational Dimensions, A Random House Media Company, 400 Bennett Cerf Drive, Westminster, MD, 21157.

Videos, Color Slides, black and white prints on weather, climate change, and related topics for teachers and educators are available from the National Center for Atmospheric Research. Call or write NCAR Information Services, P.O. Box 3000, Boulder, Colorado, 80307-3000. (303) 497-8600 or (303) 497-8606.

Computer Software

"Knowledge Tree on Global Climate Change," HyperCard stack for the Macintosh, includes two data disks and one system disk, developed by the Center for Science and Engineering Education, Lawrence Berkeley Laboratory. Available for $15 from:
Climate Protection Institute
5833 Balmoral Drive
Oakland, CA 94619

The Hole in the Sky

This kit for middle school students (grades 5–8) is based on the concepts and contents of "Science in American Life," a major exhibit at the National Museum of American History at the Smithsonian in Washington, D.C. It contains six modules that focus in various ways on the problem of the ozone layer.

Tom Snyder Productions, Inc.
80 Coolidge Hill Road
Watertown, MA 02172
(800) 342-0236

Global Warming Internet Sites

There are a multitude of Internet sites concerning global warming and the greenhouse effect, and more are being added constantly. Here are a few we found of interest.

World Resources Institute
http://www.wri.org/wri/enved

Student Conference on Global Warming
http://www.covis.nwu.edu/TeacherPointer/teachers/Gp24984

Center for Environmental Information
http://www.awa.com/nature/cei

World Climate Report
http://www.nhes.com/home

Global Warming Update: A paper by NCDC Senior Scientist Thomas R. Karl
http://www.ncdc.noaa.gov/gblwrmupd/global

IEA Greenhouse Gas R&D Programme
http://www.ieagreen.org.uk

CoVis Interschool Activity: Global Warming
http://www.covis.nwu.edu/globalWarming/global

Planetarium Program

"The Greenhouse Effect And Our Future," audio recording and art portfolio for use with a Macintosh computer and laser printer. For information, write to Star Theatre Productions, 102 Lancaster Rd., West Hartford, CT 06119.

Literature Connections

There are many excellent nonfiction books on global warming and the greenhouse effect, the issue of ozone depletion, atmospheric pollution, and the worldwide environmental crisis. Several of these are listed in the Resources section on page 146.

There are many fine science fiction works for teenagers and adults, and other literature that provides insight into these pressing issues.

We have included some books for younger students in case you are adapting some of the activities for them, or in case you think older students may enjoy them in connection with the hands-on science activities in this guide. We are sure you and your students have other favorite books on these topics and we welcome hearing about them.

The Day They Parachuted Cats on Borneo: A Drama of Ecology
by Charlotte Pomerantz; illustrated by Jose Aruego
Young Scott Books/Addison-Wesley, Reading, Mass. 1971
Grades: 4–7

This cautionary verse tale explores how spraying for mosquitoes in Borneo eventually affected the entire ecological system, from cockroaches, rats, cats, and geckoes, to the river and the farmer. "Tarapussycats" are dropped in to eat the surplus of rats resulting from the imbalance from the spraying. With its strong, humorous text, the book is successful as a dramatic reading, illustrating how "one thing leads to another" in a complex system.

The Earth is Sore: Native Americans on Nature
adapted and illustrated by Aline Amon
Atheneum, New York. 1981
Grades: 4–Adult

This collection of poems and songs by Native Americans celebrates the relationship between the Earth and all creatures and mourns the abuse of the environment.

The Endless Pavement
by Jacqueline Jackson and William Perlmutter
illustrated by Richard Cuffari
Seabury Press, New York. 1973
Grades: 5–8

Josette lives in a strange, bleak future where people are the servants of automobiles, and are ruled by the Great Computermobile. One night the "Screen" goes blank, and her father reminisces about what it was like before pavement, when there was grass, "a soft green blanket that people used to walk on." Josette is inspired and sabotages the Computermobile, starting a mass pedestrian movement.

The Faces of CETI
by Mary Caraker
Houghton Mifflin, Boston. 1991
Grades: 6–12

In this science fiction thriller, colonists from Earth form two settlements on adjoining planets of the Tau Ceti system. One colony tries to survive by dominating the natural forces they encounter, while those who land on the planet Ceti apply sound ecological principles and strive to live harmoniously in their new environment. Nonetheless, the Cetians encounter a terrible dilemma—the only edible food on the planet appears to be a species of native animals called the Hur. Two teen-age colonists risk their lives in a desperate effort to save their fellow colonists from starvation without killing the gentle Hur.

The Fire Bug Connection
by Jean Craighead George
HarperCollins, New York. 1993
Grades: 6–12

In this ecological mystery, 12-year-old Maggie receives European fire bugs for her birthday. But they fail to metamorphose and grow grossly large and explode instead. Maggie uses scientific reasoning to determine the cause of their strange death.

Just A Dream
by Chris Van Ausburg
Houghton Mifflin Co., Boston. 1990
Grades: 1–6

When he has a dream about a future Earth devastated by pollution, Walter begins to understand the importance of taking care of the environment.

The Lorax
by Dr. Seuss (Theodor S. Geisel)
Random House, New York. 1971
Grades: Preschool–8

As the beautiful forest is destroyed because it contains "truffula" trees, needed to make "thneeds," human exploitation of the environment is emphasized. The Lorax, who speaks for the environment, explains that the deforestation has affected not only Brown Bar-ba-loots who eat truffula fruits, but also the swans, fish, and other creatures. Ironically, the thneeds factory owner is placed in charge of the last truffula tree seed.

One Day in the Tropical Rain Forest
by Jean C. George; illustrated by Gary Allen
Thomas Y. Crowell. New York. 1990
Grades: 4–7

A young boy works as an assistant to a scientist as they seek a new species of butterfly for a wealthy industrialist who might preserve a section rain forest scheduled to be bull-dozed. They finally arrive at the top of the largest tree in the forest and spy the nameless butterfly.

Sweetwater
by Laurence Yep; illustrated by Julia Noonan
Harper & Row, New York. 1973
Grades: 5–8

Two different groups living on a star colony, the Mainlanders and the Silkies, attempt to save their ways of life.

Who Really Killed Cock Robin?
by Jean C. George
Harper Collins, New York. 1991
Grades: 3–7

This compelling ecological mystery examines the importance of keeping nature in balance, and provides an inspiring account of a young environmental hero who becomes a scientific detective.

Will We Miss Them?
by Alexandra Wright
Charlesbridge Books,
Watertown, Mass. 1992
Grades: K–6

A sixth-grader writes about "some amazing animals that are disappearing from the earth." For each of a dozen endan-gered species, interesting facts and illustrations are noted, ending with the question "Will we miss . . .?" A world map showing the location of each species is included. The book presents a hopeful message that we don't have to miss them, we can save them!

Assessment Suggestions

Selected Student Outcomes

1. Students are able to explain and debate the major issues that relate to global warming.
   - How the Earth's global temperature is changing during the current century.
   - How the present climate relates to climatic changes thousands of years ago.
   - The greenhouse effect and its impact.
   - How carbon dioxide creates a kind of greenhouse effect in the Earth's atmosphere.
   - How carbon dioxide can be detected and monitored.
   - The sources that contribute to increasing carbon dioxide in the atmosphere.
   - The relative amounts of greenhouse gases contributed by different countries.

2. Students are able to discuss some of the possible consequences of a warmer world and recognize the uncertainty of these predictions.

3. Students formulate their own opinions about what, if anything, should be done to slow the production of carbon dioxide and other greenhouse gases.

4. Students formulate thoughtful questions about global warming and the greenhouse effect.

Built-In Assessment Activities

What Have You Heard?

In the first activity, the students list what they have heard and the questions they have about the greenhouse effect. Their lists provide insight into the prior knowledge that students bring into the classroom. At various points throughout the unit, review the list with the students and have them comment on what ideas have changed and what questions they have answered.
(Outcomes 1–4)

A Review of the Greenhouse Effect

Sessions 2–5 explore various aspects of the greenhouse effect in the atmosphere. Each session ends with questions about that particular topic. At the end of Session 5, review the questions from previous chapters so that students can articulate the information they have learned, orally or in writing. An effective approach is to have small groups of 2–4 students discuss key questions and report back to the class.
(Outcome 1)

The Effects Wheel

In Session 7, the students create an Effects Wheel in which they speculate about a chain of cause-and-effect relationships that might occur from global warming of five degrees Fahrenheit. The teacher can examine the students' work for evidence of their understanding of the ideas presented in the previous lesson.
(Outcome 2)

World Conference

In the last session of the unit, the students conduct a World Conference. They then stop the role play to express and solidify their own ideas. At this point, ask the students to write essays to express their own opinions and formulate plans for action. The conference, the discussion, and the essays all provide the teacher with insights into the students understanding of global warming and the greenhouse effect.
(Outcomes 1–4)

Additional Assessment Idea

Global Warming Newsletter

As a class project, have your students create a newsletter that discusses what is known and what is controversial about global warming, and what all citizens need to know about this important topic.
(Outcomes 1–4)

Summary Outlines

Session 1: What Have You Heard About the Greenhouse Effect?

Before the Day of the Activity

1. Ask students to bring in articles about global warming.
2. Make copies of handout and homework sheets.

On the Day of the Activity

1. Tape butcher paper to the wall.
2. Set aside sentence strips and masking tape.
3. Organize the room for groups of students.

Writing About the Greenhouse Effect

1. Tell students you want to find out what they know about the greenhouse effect and how it may warm the Earth.
2. This is how scientists start an investigation.
3. Ask students to think about what they've heard.
4. Students have 3 minutes to make their lists.
5. Students may draw pictures, then describe in words.
6. Questions on other side of paper.

The "Mind-Swap"

1. Students work in teams of 3 or 4 to share information.
2. Explain the rules.
   a. No interruptions as one student shares.
   b. Next person says only what is not covered by previous person.
   c. When everyone has shared, teams ask questions and discuss.

Discussing What We've Heard

1. Call on each group—one thing they've heard or know.
2. Discuss disagreements.
3. During unit, class will consider accuracy of listed statements.
4. If "hole" in ozone layer listed, explain it is NOT cause of greenhouse effect.

What We Don't Know About the Greenhouse Effect

1. Groups discuss questions; each person writes one question on sentence strip.
2. Considered a question if no one in group knows answer.

3. Distribute strips of paper and marking pens.
4. Students post their questions on the wall.
5. Ask students if they can answer any of the questions posted.
6. Scientists do not know all the answers about global warming; students are learning with the scientists! Will return to questions later.

Thinking About Climate Change

1. Ask about evidence of climate change, such as where the students live.
2. Some scientists believe the average temperature of the Earth is increasing. Ask students how to measure average temperature.
3. Hand out graph on average world temperature. Ask volunteer to read scales of upper graph.
4. What happened to the temperature over the past 100 years?
5. Scientists are concerned about rising temperatures; this is already a warm period for the Earth.
6. Ask volunteer to read the scales in lower graph.
7. Ask class to describe how temperature has changed in past 450,000 years.
8. Discuss Ice Ages. When was last interglacial period?
9. Students meet in small groups to discuss: How many interglacial periods in past 100,000 years? How long into the current warm period? How much has the temperature risen in past 110 years?
10. Discuss answers with entire class.
11. Some scientists say the Earth will cool off and we'll have another Ice Age; others say the Earth will be warmer by middle of next century.

Introducing the Homework

1. Hand out "Everyone Likes To Talk About the Weather . . ."
2. Explain purpose of assignment—to interview older people about their memory of climate 50 or 60 years ago.

Session 2: Modeling the Greenhouse Effect

Before the Day of the Activity

1. Collect two 2-liter soda bottles for each team.
2. Cut three pieces of cardboard.

On the Day of the Activity

1. Practice drawing arc with 4-foot radius on chalkboard.
2. Make copies of data sheet for every two students.
3. Close doors and windows against drafts.
4. Prepare light bulbs and stands.
5. Tape thermometers and cardboard strips to insides of bottles.
6. Pour potting soil into each bottle.
7. Have one set of equipment for demonstration.

Discuss the Homework

1. Invite students to discuss answers.
2. Some scientists believe temperature data suggests the Earth's climate is changing due to human activities.
3. Explain purpose of the unit: How the greenhouse effect may be causing the Earth to warm up.

Why Do We Need a Model of the Atmosphere?

1. Since atmosphere is large and complex, it's difficult to make measurements and conduct experiments.
2. Ask class to suggest how to determine the average temperature of the Earth's atmosphere.
3. Ask what difficulties scientists might have determining whether or not the Earth is warming up.
4. The greenhouse effect and global warming are new areas of study, and scientists do not have all the answers.
5. To test theories, build a model and experiment with it.
6. Students first need a clear idea of what the atmosphere is like.

How High Does the Atmosphere Go?

1. Using a string, draw arc with 4-foot radius on the chalkboard.
2. How far would Earth's atmosphere extend above surface in the drawing?
3. Explain that 90% of the Earth's atmosphere is within the chalk line!
4. Draw the Space Shuttle 2 inches above chalk line.
5. If the Earth were an apple, the atmosphere would be as thick as an apple's skin.

A Closer Look

1. Draw a small rectangle around part of the chalk line you will "blow up" 200 times.
2. Draw a rectangle 2 feet tall, with connecting lines to the small rectangle. Label the bottom "the ground" and the top "10 miles high."

3. Ask how high is the world's tallest mountain? Draw it (5 miles) and the tallest building (1 mile), and your school.
4. The lowest 7–10 miles is troposphere; 10–30 miles is stratosphere. Both contain 99% of the air.
5. Why is atmosphere thicker (denser) at bottom? [gravity]
6. Now make a model of atmosphere—what happens when light shines on it?

The Greenhouse Experiment

1. Assemble students into groups of 3 or 4. Hold up a soda bottle, representing atmosphere, and a light bulb for the Sun.
2. Point out the thermometer in the bottle, the light bulb covered with paper.
3. One bottle is the "control." The other is covered with plastic to trap heat.
4. Teams set up experiments.
5. Show how to arrange bottles, space them equally from the light bulb with a half-inch spacer.
6. Students read temperatures in bottles; determine amount of degrees needed to "zero" the thermometers.
7. Students finish setting up experiments.
8. Student work in pairs to read and record data sheet.
9. Predict which bottle will get warmer and by how much.
10. Hand out data sheets and check equipment.
11. Start timing, with one reading per minute for 15 minutes.
12. Help teams as needed.

Analyze the Data

1. Stop experiment after 15 minutes, or when temperatures level off.
2. Pairs of students swap data, so temperatures from one "control" and one "experimental" bottle are on each data sheet.
3. Students draw lines between dots with red pen (experimental bottle) and green pen (control bottle).
4. Groups write names on sheets and teacher posts one data sheet from each group on wall.
5. One student in each group summarizes results.
6. Students summarize comparisons between control and experimental bottles.
7. Ask: Why did temperatures go up? Why did they level off? Why did closed bottles level off at higher temperature than the open bottles?

Understanding the Greenhouse Effect

1. Define equilibrium temperature. What difference did students measure between equilibrium temperatures of closed and open bottles?
2. Ask students what happens when a car is parked in the sun; does the temperature continue to increase; how you can cool off the car?
3. Trapping heat this way is called the greenhouse effect.
4. Students relate model to real atmosphere: What did the bottle and light bulb represent? How did the air inside behave? What is the difference between this experiment and the real atmosphere?
5. Scientists think something is trapping heat in the Earth's atmosphere; but it is NOT a solid barrier.

Session 3: The Global Warming Game

Before the Day of the Activity

1. Ask students to help prepare materials. Make copies of game board and cards.
2. Assemble game boards.
3. Cut out three sets of game cards for each board.
4. Using colored paper, make signs for Visible Light Photon, Infrared Photon, Carbon Dioxide Molecule, and Rock Molecule.
5. Make signs on butcher paper: Outer Space, The Atmosphere, The Earth.

On the Day of the Activity

1. Arrange desks or tables so there is an open area for the drama. Label areas: Outer Space, The Atmosphere, The Earth.
2. Push tables or desks together for teams of four.
3. Set out materials for game.
4. Tape sample game board to wall for demonstration.
5. Draw data table on board for summarizing game results.

Review the Greenhouse Experiment

1. Students summarize main ideas from previous session: Energy enters the bottle and heats it. Energy leaves the bottle. With plastic on the top, warm air is trapped inside.
2. In Earth's atmosphere there is no solid barrier. Instead, heat is trapped by "greenhouse gases."

Introducing the Global Warming Game

1. Tell students they will play a game that illustrates how greenhouse gases retrap heat in the Earth's atmosphere.
2. Organize teams into groups of four (or three) and hand out game boards.
3. Turn on light bulb and ask students how they know it is on. [light]
4. Ask volunteer to close eyes and say when bulb gets close. [feels heat]
5. These experiences demonstrate two kinds of energy packets, or photons, that come from light bulbs and from the Sun: visible light photons, and infrared heat photons.
6. Visible light photons—white beans; infrared photons—red beans.
7. Matter made of molecules that are made of smaller atoms. In liquids and gases the molecules move around, in solids they are held more together.
8. What kinds of molecules make up the air? (76% nitrogen, 22% oxygen, 0.03% carbon dioxide, and others.) This small amount of CO_2 keeps the Earth from an Ice Age. More CO_2 will make the Earth warmer.

The Interaction Between Photons and Molecules

1. Point out signs for: Outer Space, the Atmosphere, and the Earth. Note those areas on the game boards.
2. Ask for 3 volunteers: Hand the "Visible Light Photon" to student who stands in Outer Space; "Carbon Dioxide Molecule" sign to student who stands in The Atmosphere; "Rock Molecule" sign to student who stands on The Earth. They hold up their signs.
3. Student carrying visible light walks past student holding carbon dioxide, and approaches The Earth.
4. Visible light came from Sun, traveled in a straight line; now passing through Earth's atmosphere.
5. CO_2 does NOT absorb visible light.
6. The student carrying visible light stops at the Earth; photon may be absorbed or reflected.
7. If energy from photon absorbed, the visible light photon carrier hands his photon sign to rock molecule. Rock molecule absorbs this energy by vibrating. Student jiggles "Rock Molecule" sign, to show rock warming up.
8. Molecules moving in hot substances can't be seen because they are too small, but, if you touch the rock, your skin senses the vibration as warmth.

9. The rock molecule cools off after a while, and stops vibrating. It gives off an infrared photon.
10. The "rock molecule" stops vibrating and hands the "Infrared Photon" sign to student who first carried the photon sign. The student who now holds the infrared photon sign carries the energy away from the Earth.
11. Ask the infrared photon carrier to stop next to the carbon dioxide molecule on her trip away from the Earth.
12. Carbon dioxide molecules absorb infrared photons. The photon carrier hands her "photon" to the carbon dioxide molecule, who vibrates. When the carbon dioxide molecule cools off, it stops vibrating, and the infrared photon sign is handed back to the carrier.
13. The photon might go back into space, or it might head back to the ground, where it might heat up another rock or another CO_2 molecule.
14. Give students a chance to ask questions about what happens when photons of energy from the Sun interact with matter.

Rules for the Global Warming Game

1. To start, each group places 12 white beans (visible light photons) in a row along the top. They also receive two stacks of cards marked "Heads-Reflected" and "Tails-Absorbed." They will shuffle each deck separately and place them on the board in the space marked.
2. Players in each group take turns to see what happens to one photon from the time it enters the atmosphere to the time it exits into space. The players will push a photon down to the Earth, toss a coin to see if it is reflected or absorbed, and then pick a card from a pile, determined by the coin toss (heads or tails), to see what happens next.
3. If visible light photon is absorbed, it warms up the object that absorbs it. The object cools off by emitting an infrared photon. To show this, the player replaces the white bean with a red bean, and records a "W" on the score sheet.
4. Each player plays the photon until the instructions on the card say the photon goes off into space. The photon is removed from board, and the next student begins with the next photon. Play for Round 1 continues until all 12 photons removed from the board.
5. Students play the game three times. Round 1—to see what happens on an imaginary planet where there is only oxygen and nitrogen in the atmosphere. Round 2—add some CO_2 to the atmosphere, so it is more like a real planet, such as the Earth. Round 3—play game again, adding more CO_2.

6. The object of the game is to compare the amount of warming that occurs when more CO_2 is added to the atmosphere in Rounds 2 and 3.

Playing Round 1 of "The Global Warming Game"

1. A member from each group collects materials for Round 1.
2. As the groups finish setting up, remind them of the rules: "Push a Photon—Toss a Coin—Pick a Card."
3. Check to see all groups are playing correctly.

Playing Rounds Two and Three

1. When all groups finish Round 1, ask how many times absorption of photons warmed the Earth—fill this number in for each group on the chalkboard data table.
2. Announce second round. Introduce CO_2 to the atmosphere. Hand out "Carbon Dioxide" cards for students to shuffle and place on the game boards.
3. Use the posted copy of the game board to show the placement of three pennies randomly across the board, and how to trace around them in pencil, to represent three CO_2 molecules.
4. Show how a visible light photon passes through a CO_2 molecule on its way to Earth, or on its way to space after being reflected. Visible light photons are only rarely absorbed by CO_2, so in this game, visible light will always pass through CO_2 molecules.
5. Ask students to recall what happens when visible light photon is absorbed by the Earth. [It warms the Earth, and is emitted as an infrared photon.] If an infrared photon encounters a CO_2 molecule, it is always absorbed. When this happens, the player writes "W" on the score sheet and picks up a card from the "Carbon Dioxide" pile.
6. An infrared photon runs into a CO_2 molecule and is absorbed when you slide the exact middle of the bean sideways up the line and any part of it touches a circle that represents a CO_2 molecule.
7. CO_2 molecules cool off the way other molecules do—they give off an infrared photon. Sometimes these photons go into space, sometimes back toward Earth, where they are always absorbed. This is shown by picking a card from the "Tails-Absorbed" pile, without having to flip a coin.
8. Remind students to place the score sheet for Round 2 on their game boards, and trace in three CO_2 molecules.
9. After Round 2, students draw in three more CO_2 molecules and play the game again with the score sheet for Round 3.

10. As the groups finish data collection for Rounds 2 and 3, students record data on the chalkboard, writing in the total number of W's for each round.
11. Groups who finish early can play one more round with a mixture of carbon dioxide and methane in the "atmosphere." Use a quarter to trace a larger circle over three of the CO_2 circles to represent methane, which absorbs infrared photons more easily than CO_2.

Summarizing the Results

1. When all groups have completed at least Round 3, and entered their results, collect all game boards and equipment, and discuss the data.
2. Ask the class to suggest any trends. [Warming without CO_2 will in general be lower than warming with low CO_2, which in turn will be lower than warming in an atmosphere with a lot of CO_2. Individual groups may have different results due to statistical variations.]
3. Ask the class to suggest how the results might be summarized to see trends, or patterns in the data, more clearly. [Add up the columns to see how more photons behave.]
4. The rules of the game not simply made up, but actually show how CO_2 molecules interact with visible light and infrared photons. Scientists use more precise, but similar, games, usually on a computer, to predict what will happen when more CO_2 is added, which is called a simulation game.
5. "How does CO_2 in the atmosphere warm the Earth?" [Molecules of CO_2 absorb some of the escaping infrared photons, warming the atmosphere. When these molecules cool, they give off infrared photons, some of which go back to the Earth's surface. Some photons are absorbed several times before they go back into space.]
6. If students have difficulty making generalizations, ask focused questions.
7. CO_2 is called a "greenhouse gas" because it has the effect of trapping heat energy, or infrared photons, in the Earth's atmosphere, If it were not for the small amount of CO_2 in the atmosphere, the Earth would be in a continual Ice Age. CO_2 helps maintain the Earth's temperature at a comfortably high level. But a substantial increase in CO_2 is predicted by many scientists to make the Earth warmer than it is now.

8. Ask students to explain how this effect differs from the greenhouse effect they experimented with in Session 2. [The plastic bottle greenhouses heated up—a physical barrier stopped the warm air from mixing with the cool air. In the game, CO_2 gas traps infrared photons so the heat energy is returned by many more molecules before leaving the Earth.]
9. Emphasize that life on Earth has always depended on the greenhouse effect. The concern now is with the rapid increase in CO_2 due to human activity.
10. Other gases in the atmosphere absorb infrared photons even more readily than CO_2. These gases, which include methane, nitrous oxide, and chlorofluorocarbons (CFCs), are also increasing rapidly due to human activity.
11. Gases such as methane and CFCs can be represented by larger circles in the game. What would happen if students played another round with larger circles? [There would be more W's on the score sheet—even more heat would be trapped.] If some of the groups played an additional round with larger circles for methane, ask them to report their results.
12. CO_2 is considered the biggest problem because so much of it produced every year.

Homework: The Past 160,000 Years

1. Hand out homework sheet.
2. Discuss graphs on page 57 so students are able to interpret them.
3. Tell students they will perform lab experiments to determine amount of CO_2 in different samples, such as the air in the classroom, and in their own breath.

Session 4: Detecting Carbon Dioxide

Several Weeks Before Beginning this Unit

1. Collect glass wine bottles.
2. Obtain bromothymol blue (BTB) solution.

Before the Day of the Activity

1. Make a one gallon solution of BTB and water. For each group of four students, fill a dropper or squirt bottle with about 6 oz. of prepared BTB solution.
2. Make one copy of the data sheet and homework sheet for each student, and one copy of observation sheet for each group of four students.

3. If necessary, make two measuring cups for each group.
4. Read through the instructions and practice generating CO_2 gas and collecting it in a balloon; using the gas to put out a candle flame; and detecting CO_2 with BTB.

On The Day of the Activity

1. Provide each group of four students with a half cup of baking soda and two-thirds cup of vinegar.
2. Assemble equipment for each group on a tray. Use one set for demonstration.
3. Secure the candle in the middle of the container with drops of wax. Assemble all equipment needed for a demonstration.
4. Have an air pump available for passing from group to group.

Review

1. Ask the class to recall the results of the game by asking questions focusing on each round.
2. Discuss the answers to questions on the homework sheet.
3. Students will probably conclude that the amount of CO_2 determines average global temperature. Point out that scientists don't know for certain if changes in CO_2 cause changes in temperature, or the other way around.
4. Ask what the range in temperature in Antarctica has been over the past 160,000 years. [about 23°F] Temperature differences tend to be more extreme at poles and less near the Equator. Earth's average temperature now is about 59°F. It was 9°F or 10°F colder during the Ice Ages.
5. Ever since the Earth has had an atmosphere, the greenhouse effect has warmed the Earth. Over past 100 years, the concentration of CO_2 has increased dramatically, and average global temperature has increased by about 1°F. Many researchers suspect the increase in CO_2 caused the increase in temperature, but it is still a scientific debate. Most researchers believe that, if we do not change the rate at which we burn fossil fuels, the Earth will be from 2°F to 7'°F warmer by the middle of the next century.

Demonstration of Carbon Dioxide's Properties

1. Ask the class what they already know about CO_2 gas: what it looks like, where it is produced, and its properties.
2. To detect CO_2 gas, a good supply is needed. Does anyone know an easy way of making CO_2? [Breathing is an easy way, but doesn't produce pure CO_2.]
3. An easy way is in a chemical reaction: vinegar and baking soda.

4. Demonstrate, using an empty bottle. Measure 3½ oz. (100 ml) of vinegar and pour it into a bottle. Make a funnel from paper to pour the vinegar in with the baking soda.
5. Have a volunteer light the candle in the container, then stand nearby. Put four heaping teaspoons of baking soda into the bottle. Ask the volunteer to observe and describe, then place her finger over the end of bottle for a few seconds and report what it feels like.
6. Ask the class to watch what happens as you hold the opening of the bottle directly above the candle flame, and "pour" the gas (not the liquid) onto the flame. [The CO_2 fills up the container, blocks oxygen from getting to the candle, so the flame goes out.]
7. Empty out the bottle for use by a student group.

Collecting a Sample of Carbon Dioxide

1. Students prepare a balloon of CO_2 gas. Pour 3½ oz. (100 ml) vinegar into a bottle, add 4 heaping teaspoonfuls of baking soda. Wait for less than one second to allow the CO_2 to drive the air out, stretch the neck of a balloon over the opening, and inflate to about 4 inches in diameter. Tightly tie off the neck of the balloon with twist-tie.
2. Show how to use the air pump to fill a second balloon to the same size, and close with twist-tie.
3. Have one student from each group pick up materials. Pass the air pump to the first group, and have all groups begin.

Detecting Carbon Dioxide

1. As students work, prepare the next demonstration by measuring out half an ounce (15 ml) BTB into three plastic cups.
2. When the students finish, demonstrate how to test for CO_2. Insert a straw into the neck of a balloon and seal it with second twist-tie. Slowly loosen the top twist-tie, releasing gas until the balloon just passes through the hole in a tape roll. Clamp off the flow of gas with your fingers; place the bottom end of the straw into the BTB. Loosening grip, slowly bubble gas through the blue liquid. Squeeze the balloon to release all the gas and observe the color of the liquid.
3. Hold up an Observation Sheet and Data Sheet, and explain how results should be recorded.
4. Point out safety considerations. While none of these chemicals are dangerous, always take safety precautions.
5. Students adjust the size of the balloons and bubble the gas samples through the BTB.

6. Ask, "How does BTB react in the presence of almost pure CO_2?" [It turns yellow.]
7. BTB stands for bromothymol blue solution, used to test for the presence of CO_2 gas. Bright yellow indicates almost pure CO_2; green, some CO_2; blue-green, a little CO_2; and blue, no detectable CO_2.
8. Students record the results on their data sheets; writing a descriptive sentence or two.
9. Students put equipment back on the trays and return.

Discussing Results and Assigning Homework

1. Ask students to compare how the two gas samples affected BTB. Did the room air affect the BTB at all? [Results depend on BTB concentration, and ventilation—in most cases, room air does not change the color of BTB very much]
2. Ask students what this means in terms of the concentration of CO_2 in air. [The concentration of CO_2 in air is near the limit of the sensitivity of the BTB test. CO_2 makes up only 0.0355% of the atmosphere.]
3. Remind the class of links between CO_2 levels in the atmosphere and global temperature. Ask about sources of CO_2 in the atmosphere. List them on the chalkboard or on butcher paper.
4. In the next session students will use same method to compare amounts of CO_2 in human breath and car exhaust.
5. Hand out homework assignment.

Session 5: Sources Of CO_2 in the Atmosphere

Before the Day of the Activity

1. Use a car with a round exhaust pipe. Prepare a cone for collecting CO_2 from a manila folder. Approach the car's exhaust pipe from the side and hold your breath so you do not inhale any gases.
2. Make one copy of two-sheet homework assignment and data sheet for each student, and one copy of the observation sheet for every group of four students.

On the Day of the Activity

1. Park the car within close walking distance of the classroom.
2. Assemble equipment for each team.
3. Have an air pump for passing from group to group.

Review

1. Discuss answers to the homework questions.
2. Review results from previous session. [Blue BTB turned green, then yellow as CO_2 bubbled through it; the level of CO_2 was not very high in air.] On the chalkboard, draw the scale from the previous session showing how BTB changes color when exposed to different amounts of CO_2.

Collecting Samples of Car Exhaust

1. The class will compare amount of CO_2 in car exhaust with other samples, using BTB. There will be brief "field trip" to the car.
2. Organize the class into groups, each with a twist-tie and balloon. Take extra balloons and twist-ties.
3. At the car, tell students to work together to tie the neck of the balloon tightly with a twist-tie.
4. Collect each sample in turn. Collect two extra samples in case of loss. Return to classroom.

Conducting the Experiment

1. Four gas samples to be tested. List:
   1) car exhaust
   2) air from room
   3) CO_2 from vinegar and baking soda
   4) human breath
2. Students have car exhaust samples, have learned how to collect the air sample with air pump, and how to make/collect pure CO_2 using vinegar and baking soda.
3. Ask the class how they should collect the human breath sample? [By blowing up the balloon]
4. After collecting all four gas samples, make the samples the same size, as before, by adjusting each balloon so it passes through the hole in a roll of tape.
5. Hand out data sheets and summarize the procedure for using BTB.
6. Students record which gas sample is put into which color balloon. After they do each test, they record the color of the BTB in the space at the bottom.
7. Students keep the BTB solution in cups until the end of the experiment, to compare the colors. When finished, they should fill in the names of the four samples along the line on the data sheets, from least to most CO_2.
8. Distribute equipment trays/bottles, pass the air pump, and have students begin.

9. Help as needed. Make sure no baking soda gets into the BTB tests (baking soda neutralizes carbonic acid, preventing BTB from changing color). Students should view BTB samples against the white paper of the observation sheet to get true color results.
10. As students finish, remind them to fill in final section of the data sheet, expressing their findings in writing.

Discussing the Results

1. When all groups finish, set the equipment aside.
2. Which gas sample had the highest CO_2 content? Using the scale on the chalkboard, record the results. Fill in other samples the same way. If there are conflicting results, list each gas sample and write the number of groups with each result.
3. Ask students what the data shows about the amount of CO2 in equal volume samples of gas from these sources. [Most groups find from least to most CO_2, is: air, human breath, auto exhaust, and almost pure CO_2 from the baking soda and vinegar mixture.]
4. What additional information is needed to determine relative amounts of CO_2 contributed to the atmosphere by humans breathing, and humans driving cars? [Number of humans versus number of cars, how much time cars are driven, volume per minute of gas exhaled by humans and in car exhaust.]
5. Which of the sources of CO_2 students measured (car exhaust and human breath) could be reduced? How might this be done?
6. Remind students that some CO_2 is needed in the atmosphere, or the Earth would freeze. The problem is one of balance. Many researchers think too much CO_2 is being added to the atmosphere from burning fossil fuels such as gasoline, natural gas, coal, and oil. Each year people burn fossil fuels that required about 1 million years to form. Discuss sources of CO_2.
7. Hand out the homework assignment and discuss the bar graphs in it to make sure students can interpret them.

Session 6: Changes On Noua's Island

1. Photocopy homework articles and other masters.
2. Plan how to organize tables or desks so small groups can meet together during the session.

Where Does the World's CO_2 Come From?

Discuss the homework questions.

The Story of Noua's Island

1. Ask students to imagine themselves in a developing country threatened by increased sea levels.
2. Read the story to help students imagine such a place, and how people there may be directly affected by global warming.
3. Read the story slowly and carefully. Where indicated, stop to ask questions.
4. After the story, ask: "What would you do in Noua's situation? How might other countries respond?

Planning the International Conference

1. Announce that the students will conduct an imaginary World Conference to discuss how people might cooperate to decrease the greenhouse effect, or cope with the changes it causes. Student groups represent: Automobile Manufacturers, Island Nations, Agriculturists, Conservationists, Wood and Paper Producers. Designate the groups.
2. Organize tables or desks so students can meet in the five small groups. Give out copies of interest group descriptions.
3. Briefly describe the interest groups—at the World Conference each of the groups will present points of view, and look for possible solutions acceptable to some or all of the others.
4. It will not be a debate—everyone is affected by global warming, so no one is really "for" it. Instead, the task is to find out who is affected, how, and work out ways of doing something about it.
5. Groups appoint a reader and a recorder, and have 10–20 minutes to read the handouts, discuss, and record their ideas.
6. Five minutes before end of the session, ask for questions to be listed on each group's notes. There will be more time in the next session for conference planning.
7. Collect writing from each group for your comments.
8. Hand out reading assignment.

Session 7: The Worldwide Effects of Climate Change

1. Photocopy the Effects Wheel.
2. Read student notes from Session 6. Make written comments.
3. Photocopy "Flash! Messages," cut apart for distribution to the groups.

The Effects Wheel Activity

1. Have students recall predictions from their homework reading about possible effects of global warming, and make a list. If some effects follow from others, indicate this way: polar caps melt → sea level rises → coastal areas flooded.
2. Point out some effects are a result of other changes before them—there is a "chain reaction."
3. One way of thinking about long-term effects, like a 5°F rise in the world's average temperature, is to make an Effects Wheel.
4. Students work in same groups as previous session, so they can choose effects important to their World Conference interest group.
5. Hold up example of the Effects Wheel. Each group should write "+ 5°F" in the center circle.
6. Each group should choose four primary effects of global warming from those listed on the board and write each of these in the areas next to the inner circle, for example, "a rise in sea level."
7. Point to the next outer segments of the circle. Point out where to write the effects that would result from the sea level rise. Effects stemming from those events would be written in the next outer circle, and so on.
8. Hold up a packet of Post-Its and explain their use, and that they allow the group to write a number of ideas at once, and to make changes.
9. Remind students that in choosing primary effects they may wish to consider those of interest to people they represent in the conference.
10. Organize class into groups, hand out one blank Effects Wheel, Post-Its, pencils, and have them begin.
11. Circulate as needed. After 15 minutes, give a red pen and a green pen to each group. For effects listed on outer circle, they should circle positive outcomes in green, negative in red, neutral in black, and those with both positive and negative outcomes with both red and green.

Sharing Completed Effects Wheels

1. After about 20 minutes, ask each group to report one of their primary effects, and one positive and one negative outcome of it.
2. Ask each group to report how many positive, negative, and neutral outcomes resulted. Make a table as the groups report these. Ask class to decide whether, on the basis of their predictions, the outcomes would be mostly negative, mostly positive, or a combination.

3. Each group posts its Effects Wheel on the classroom wall. Invite students to look at all the Effects Wheels for ideas.

Preparing for the World Conference

1. Return each group's notes from the previous session, with your comments. Tell them to go over notes, adding ideas, and referring to their Effects Wheels.
2. Explain that each group should be ready to state their position, ask questions or make statements about other groups' positions, and suggest solutions.
3. The main purpose of the conference is not to find someone to blame for global warming, but to find out who and what is affected, and what can be done by acting together.
4. Hand out the "Flash! Messages" slips. Students do not need to use any of the ideas.
5. During the small group discussions, circulate to offer suggestions.
6. Five minutes before session's end, remind students they should have clear answers to all of the questions, and should decide who will speak when. Encourage students to take turns, rather than having just one student present the group's ideas.
7. Collect the preparation notes.

Session 8: World Conference On Global Warming

1. Read over the student notes on possible solutions and comment.
2. Organize space on the wall to post solution slips.
3. Organize seating in conference-style arrangement, circle or semicircle.
4. You may want to hang a banner announcing the "World Conference on Global Warming" and/or make name cards for the groups.

Begin the Conference (10 minutes)

1. Invite the students to be seated in their interest groups for the conference. Hand out notes from the previous session.
2. Introduce the conference.
3. After each group has introduced itself, give them five minutes to finalize their statements.
4. Have the groups to write each solution on a separate sentence strip.

5. Five sentence strips to each group. After five minutes, call students back together for the conference.

Interest Groups Make Presentations (25 minutes)

1. Explain the procedure, with three minutes for a group's statements and solutions, two minutes for questions and comments from other groups, and one minute for response. Stress courtesy and cooperative problem solving.
2. As solutions are suggested, post them on the board or on butcher paper.
3. After all the groups have presented, thank each group. Point out that each group has a different approach to global warming, but you are confident the conference will reach some agreement.

Discuss Possible Solutions (10 minutes)

1. Ask the delegates whether any of the solutions posted are acceptable to all or most interest groups.
2. Discuss the relative merits of these solutions.
3. Give groups 30 seconds to discuss whether people they represent would vote "Yes," "Maybe," or "No." Take one vote from each group; record the results next to the solution.
4. "Winning solutions" are those agreed on by a substantial number of groups.
5. The conference is at an end. "Leave behind your point of view as a member of an interest group and look at the solutions before us as an individual. What do you personally think about the main solutions suggested?" Take a second vote on each of the solutions.
6. Vote on each solution.
   Strongly agree—both hands up.
   Maybe a good idea—one hand up.
   Not sure—fold arms.
   Bad idea—thumbs down.
7. Summarize the class response to each solution.
8. A majority vote on issues of the environment may not necessarily be the best way, and is definitely not the only way of making decisions. There are many ways in which issues arise and can be resolved.
9. Remind the class, regardless of what we might want to do, our environment has ultimate veto power—it may not want to do what we vote it should!